THE US M3/M3A1 SUBMACHINE GUN

The Complete History of America's Famed "Grease Gun"

★ Michael Heidler ★

SCHIFFER MILITARY
4880 Lower Valley Road Atglen, PA 19310

Library of Congress Control Number: 2020930851

Edited by Bob Biondi
Designed by Jack Chappell
Cover design by Jack Chappell
Type set in Nimitz/Bodoni/Univers

ISBN: 978-0-7643-6058-9
Printed in India

Published by Schiffer Publishing, Ltd.
4880 Lower Valley Road
Atglen, PA 19310
Phone: (610) 593-1777; Fax: (610) 593-2002
E-mail: Info@schifferbooks.com
Web: www.schifferbooks.com

Opposite Page: Rex Allers with his Grease Gun in Vietnam with the 11th Infantry Brigade "Jungle Warriors," 1968. *Courtesy of Rex Allers / Richard Warren*

CONTENTS

ACKNOWLEDGMENTS

DURING MY WORK ON this book, I was fortunate to be in touch with many helpful people around the globe. Without the transfer of material and information, a book like this would not have been possible.

A very special thanks goes to a longtime friend from Pennsylvania, Frank Iannamico. As a successful author of his own numerous gun-related books, such as ***American Thunder: The Military Thompson Submachine Guns*** and ***AK-47: The Grim Reaper***, he supported this work with information, documents, and photos from his extensive archive.

I would also like to thank Jonathan Ferguson from the National Firearms Centre (Great Britain), and Richard Colton and Alex MacKenzie from the Springfield Armory Museum (USA) for their support.

It was a great pleasure for me to receive a complete set of M3 production documents from the Rock Island Arsenal Museum (USA). Many thanks to the director, Kris G. Leinicke.

The following individuals, companies, and institutions have all contributed to the success of this book. A big thanks to all of them!

David R. Albert, USA
Rex Allers, USA
George Anderson, USA
Steven Baird (Nexteer Automotive / former Saginaw Steering Gear), USA
Robin Bartel, USA
Gaston Binnerbini, Argentina
Jim Bolen, USA
Kelly Chak-Marx, Hong Kong
Richard Colton, USA (Springfield Armory Museum)
Eduardo J. Coppini, Argentina
Mike Ellis, USA
Leszek Erenfeicht, Poland
Jonathan Ferguson, Great Britain (NFC Leeds)
Enrique Javier Fernández, Argentina
Lindsay T. Fincher, USA (www.lindsayfincher.com)
Roland Gaul, Luxemburg (Musée National d'Histoire Militaire)
Anabela Giuliano, Argentina
Frank Goepfert, USA (www.atfmachinegun.com)
Luc Guillou, France
Gregory Hagge, USA (Ordnance Museum Fort Lee)
Guy Hildebrand, USA
Ken Hinks, USA
Po-Wei Huang, Taiwan
Frank Iannamico, USA
Les Jensen, USA (West Point Museum)
Brigadegeneral Panagiotis Kaperonis, Greece
Kris G. Leinicke, USA (Rock Island Arsenal Museum)
Peter Lloyd, USA (www.peteralanlloyd.com)
Alex MacKenzie, USA (Springfield Armory Museum)
Bas Martens (***SAM-Wapenmagazine***), The Netherlands
John Moss, USA
Frank C. Motley, USA
Ioannis Mylonas, Greece
Tade (Chip) Orr, USA
Karl-Heinz Pfarr, Argentina
Corey A. Shaffer, USA (www.thompsonmachine.net)
Jan Skramoušský, Czech Republic (Military Museum Prague)
Paul Smith, USA
Mike Stannard, USA
Ian Phillips, USA
Michael J. Tahirak, USA
R. D. I. Weigt, Germany
Ferdinand Weiss, Luxembourg
Herbert Woodend (+), Great Britain (MOD Pattern Room, Nottingham)
William H. Woodin (+), USA

US Army Ordnance Museum, Aberdeen Proving Ground (APG), USA
US Army Ordnance Corps / Ordnance Museum, Fort Lee, Virginia, USA
US Army Heritage & Education Center, Carlisle, Pennsylvania, USA
Airborne Assault Museum / Imperial War Museum, Duxford, Great Britain (the former Parachute Regiment & Airborne Forces Museum Aldershot)
Anderson Public Library / Indiana Room, Anderson, Indiana, USA
Armed Forces Museum, Taipei, Taiwan
ATF National Working Firearms Reference Collection, USA
Australian War Memorial, Canberra, Australia
Bundesministerium des Inneren, Austria
Deutsches Waffen-Journal, Germany
Fábrica Militar de Armas Portátiles "Domingo Matheu," Argentina
Fabricaciones Militares, Buenos Aires, Argentina
General Motors Heritage Center, Sterling Heights, USA
Hellenic War Museum, Athens, Greece
Ithaca Gun Company, USA

Military Museum Prague, Czech Republic
MOD Pattern Room, Great Britain
National Archives and Records Administration (NARA), Washington, USA
National Guard Militia Museum of New Jersey, USA
National Firearms Centre (NFC), Leeds, Great Britain
(former Ministry of Defence Pattern Room, Nottingham, Great Britain)
Musée National d'Histoire Militaire, Diekirch, Luxembourg
Museo de Armas de la Nación, Buenos Aires, Argentina
National Museum of the Marine Corps, Quantico, Virginia, USA
National Rifle Association (NRA) Museum, Fairfax, Virginia, USA
Rock Island Arsenal Museum, USA
Rock Island Auction Company, USA (www.rockislandauction.com)
Royal Armouries, Leeds, Great Britain
Small Arms Defense Journal, USA
Small Arms Review, USA
Springfield Armory National Historic Site / US National Park Service, USA
The Military Museum of the Chinese People's Revolution, Beijing, China
VISIER: Das internationale Waffenmagazin, Germany
West Point Museum, USA
Woodin Laboratory, Tucson, USA

01 INTRODUCTION

CONTRARY TO ALL HOPES, World War I was not "the war to end all wars," as the British contemporaries proclaimed so emphatically after Germany's defeat. The great world fire had indeed destroyed some old orders and empires, but had also planted the seed for new conflicts throughout the world. In Europe, arms were soon taken up again—from small border skirmishes through to the Spanish Civil War. And in September 1939, a new war of unimaginable dimensions began with the German invasion of Poland.

The initial German successes were remarkable; the Wehrmacht seemed to be unstoppable. After just a few months, large parts of Europe were under German control. The United States looked worriedly at the ominous developments in Europe, particularly in Great Britain, which already stood with its back to the wall. Despite the hesitant attitude of the American people, the government began to prepare for a coming involvement in the war. The United States entered World War II on December 8, 1941, with the declaration of war on the Empire of Japan after its attack on Pearl Harbor the day before. A few days later, on December 11, Germany and Italy declared war on the United States. So, the United States of America was back at war again. The defense industry went into full swing to meet the needs of the US armed forces and, furthermore, their allies. The production output of weapons, ammunition, and equipment was gigantic. But since a quick end of the war became more and more unlikely and increasing supplies were needed, the US had to think about a more efficient production.

The name-giver: a grease gun for automobiles, here promoted by Midland Parts & Bearings from Irving

The author is having fun with an M3A1. Thanks to the low rate of fire, the weapon shoots very pleasantly and even single shots are possible. Note the ejected case in the air.

Back in 1939, in wise foresight, the Ordnance Department had made first preparations in the search for a new submachine gun, which was to be simple, cheap, and quick to manufacture. The adopted Thompson submachine guns were of excellent quality but were not suitable for mass production on such a scale. The trials took a lot of time, and at the end, a nondescript weapon was conceived: welded together from two formed sheet-metal halves, with a simple blowback mechanism inside and a low rate of fire, the "T-20" won against all odds.

On January 11, 1943, the new weapon with the designation "U.S. Submachine Gun, Caliber .45, M3" was officially adopted by the armed forces, followed by the even more simplified model M3A1 at the end of 1944. At first, many soldiers mockingly called it "Grease Gun," owing to its visual similarity to the mechanic's tool, but it soon gained a reputation. This "Ugly Duckling" saw action not only in World War II, but also in Korea, Vietnam, and various other conflicts around the globe. China and Argentina made copies for their armies, and other countries received arms from the United States, some of which are still in active service today.

This book tells the story of a weapon that was, in spite of all its successes, always overshadowed by the Thompson models.

EARLY AMERICAN SUBMACHINE GUNS

IN AUGUST 1920, A contract for the production of 15,000 weapons of the model 1921 was signed between the Colt's Patent Firearms Manufacturing Company in Hartford, Connecticut, and the Auto-Ordnance Corporation. In the 1920s, John Thompson succeeded in selling some of his .45-caliber submachine guns to the police. But additional sales progressed very slowly. Thompson also wanted to sell these expensive weapons to the Navy, but the high rate of fire (800 rounds per minute) met with little enthusiasm. By changes in the design, the rate of fire was reduced to only 500 rounds per minute. The weapon was then accepted, but it wasn't the major order that was hoped for: the Navy acquired only 500 pieces of the "U.S. Navy Model of 1928." Some 1,000 weapons were delivered to the Army, some more to the FBI, the police, and various prisons. Several hundred weapons were sold to foreign countries, including detours to the Irish IRA, or ended up in the hands of criminals. When John Thompson withdrew from business in 1928, the company was led by his son Marcellus. But business was struggling to stay solvent. In 1939, when Auto-Ordnance was faced with bankruptcy, the entrepreneur Russell Maguire took the opportunity and acquired the ailing company. He hoped for an entry of the United States into the war—and thus for good for the weapons business. There were still more than 4,000 unsold submachine guns in stock, which he sold to the Allies for substantial money. For the series production of the slightly revised model 1928A1, the company's own capacity was not sufficient. And because the Colt's Patent Firearms Manufacturing Company was already busy at this time with other orders, the production was given to Savage Arms in Utica, New York. The Savage Arms engineers simplified the Thompson, and in April 1942, the new model was adopted as the M1. The wooden stock was no longer removable, the drum magazines could no longer be used, and the cocking handle was placed on the right side. The modified bolt assembly and other design simplifications, such as the barrel without cooling ribs, accelerated the production considerably. By eliminating the replaceable firing pin (it was now milled into the bolt face), the model M1A1 was created, which was then adopted in October 1942. By the end of production in February 1944, both companies together had delivered more than 1.5 million Thompson submachine guns.

Thompson Model 1921 in a case as used by the FBI. *Courtesy of the Rock Island Auction Company*

Thompson US Navy Model of 1928. *Courtesy of the Rock Island Auction Company*

Manufacturer's markings on the receiver. Note the overstamped date: 1921. *Courtesy of the Rock Island Auction Company*

The simplified model M1A1 was adopted in 1942. *Courtesy of the Rock Island Auction Company*

An M1A1 in action on Okinawa in May 1945. The other soldier carries a BAR (Browning Automatic Rifle).

Reisuing Model 50 in .45 ACP caliber. *Courtesy of the Rock Island Auction Company*

The Harrington & Richardson Arms Company was the sole manufacturer of the Reising. *Courtesy of the Rock Island Auction Company*

The Reising was designed as a select-fire weapon. *Courtesy of the Rock Island Auction Company*

The case was popular among agencies that used the weapon only sporadically. *Courtesy of the Rock Island Auction Company*

The Reising Model 55 was shorter, due to the missing muzzle brake and the wire folding stock. *Courtesy of the Rock Island Auction Company*

Owners manual for the "H&R Reising Submachine Gun"

Testing of the M2 at Aberdeen Proving Ground. In the course of war more and more tasks were taken over by women.

The weak cartridge of the M2 often did not show the desired effect on the target.

Markings on an M2 carbine. Note the fire selector on the left.

Most of the M3s were used by the American and British forces in Europe. Large numbers were also shipped as aid to the Soviet Union. In contrast, the weapons were quite scarce in the Pacific theater because the US Marines used mainly the Thompson models. When looking for a more effective submachine gun, the choice of the Marines fell on the M50 Reising (wooden stock) and M55 (with folding stock and without muzzle brake) in .45 caliber, which was adopted in 1942. The weapon was developed by Eugene Reising and manufactured exclusively by the Harrington & Richardson Arms Company in Worcester, Massachusetts. Although time consuming and expensive to manufacture, the Reising soon acquired a bad reputation. In the rush, in-depth testing was foregone—and now the weapons failed under combat conditions. Perfectly cleaned and lubricated, it worked well, but the Reising's close tolerance and delicate magazine proved unreliable in sand and mud. The need to lubricate all moving parts intensified the problem even more. Thanks to the design of firing from the closed bolt, the accuracy in semiautomatic was very good. When it was firing bursts, it was hard to keep the lightweight Reising on target. It wasn't long before most weapons were handed over to guard units; many others were given to police departments. By the end of production in 1945, approximately 114,000 pieces of both models together had been manufactured.

Although not a submachine gun by definition, the M2 carbine as a select-fire weapon could be used in a similar manner. It was a modified version of the semiautomatic M1 carbine (.30 carbine caliber) and adopted in early 1945. On the one hand, the M2 was popular because of its light weight, but on the other hand, it was the light weight that made aiming difficult when firing bursts. In addition, the weak cartridge did not always show the desired effect on the target.

Another submachine gun that was being evaluated by the Army was the Hyde-Inland M2. Developed by the Inland Manufacturing Company, the weapon, in .45 caliber, performed well in the tests at Aberdeen and had been proposed by the Ordnance Department as the new "U.S. Submachine Gun M2." For production, the Marlin Firearms Company in North Haven, Connecticut, was awarded a contract for 164,450 weapons. The first weapons were to be delivered in December 1942. However, technical problems delayed the start of series production, so that by the end of March 1943, not a single weapon had been manufactured. In particular, the production of six parts that were pressed out of powdered metal were failing, causing great trouble. In endurance tests, two of these pressings broke. Meanwhile, the Ordnance Department had continued the tests with other weapons and ultimately chose the T-20 (M3). When production of the M2 ceased, fewer than 500 weapons were produced.

The unfortunate Hyde-Inland M2. *Courtesy of the Springfield Armory NHS / US NPS*

The wooden stock and the milled parts contradicted the goal of simple manufacturing. *Courtesy of the Springfield Armory NHS / US NPS*

03 SEARCHING FOR A NEW SUBMACHINE GUN

AFTER THE FIRST BATTLES of World War II, the United States recognized that the submachine gun had become an indispensable weapon of the infantry. Compact and handy, with a high rate of fire and sufficient magazine capacity at short distances, it was superior to the carbines and rifles. Therefore, in 1939, the Ordnance Department made first preparations to search for a new submachine gun to replace the expensive and complicated-to-manufacture Thompson models. Later in the war, some foreign models such as the German MP40 and the British Sten showed that weapons of simplified design also could prove effective in combat.

At that time, such invitations to tender were executed as follows: on the basis of rough requirements put together by the Ordnance Department, various companies and individuals developed the appropriate weapons. These were then compared in a series of tests with each other by the Army. The winner could earn a very lucrative defense contract. The losers, however, left empty-handed. Often, weapons were rejected due to minor imperfections, without submitting proposals to the manufacturer on how to improve his weapon in order to meet the requirements. Once rejected, there was little opportunity for the manufacturer to revise his weapon and submit it again. The Ordnance Department was extremely passive, and there was no real cooperation with the designers. Nevertheless, plenty of domestic and foreign participants were always found that spared no financial expenditure in the hope of a large order—and thereby certainly could face financial ruin.

In 1939, the Ordnance Department asked the Army to prepare a list of requirements for a new submachine gun. The result was two lists, because the infantry and cavalry had a few different ideas:

	Infantry Requirements	Cavalry Requirements
Caliber:	.30 or .45	.30 or .45
Cooling	air	air
Weight	7 to 10 pounds	maximum 11 pounds
Length (stock extended)	36 inches	34 inches
Length (stock folded)	23 inches	not essential
Sights	open	open, adjustable
Magazine capacity	50 rounds	20 and 50 rounds
Rate of fire	500 rounds per minute	600 rounds per minute
Type of fire	automatic and semiautomatic	automatic and semiautomatic
Grip	pistol	pistol
Sling	yes	yes
Maximum effective range:	300 yards	—

After the Ordnance Department's consultation with Aberdeen Proving Ground (APG), some suggestions were made. Aberdeen was the oldest research and testing facility of the US Army, with many years of experience in the development and testing of weapons. The facility was established during World War I to analyze foreign weapons and to support the development of national arms. The facility was located on a peninsula in the Chesapeake Bay, not far from the town of Aberdeen, Maryland. The 30,000-acre site offered the best conditions for vehicle test driving and shooting tests. The official opening of the APG was held on October 20, 1917.

Aberdeen Proving Ground in 1942, with the Ordnance School in the foreground. *APG*

The bolt of the new weapon was to be made as heavy as necessary to keep the cyclic rate below 500 rounds per minute. Fired from a standing position in full-automatic mode, the weapon should achieve a hit rate of 90 percent on a 6-by-6-foot target at a distance of 50 yards. Changing the magazine should be possible whether the bolt was in the open or closed position. During the testing, it was decided that a switch for semiautomatic functioning was not required.

Rapid manufacturing of the potential new submachine gun was of great importance. The United States at that time was on the brink of entering the war—it was only a matter of time. From the start, submachine guns that needed extensive machining effort or contained complicated mechanics were eliminated. The demand was for an all-metal weapon made of stamped sheet-metal parts. Also, no critical metals, such as aluminum, were to be used. Even the requirement for a carrying sling and swivels on the weapon was canceled.

THE FIRST PHASE OF THE COMPETITION, 1939–42

The weapons taking part in the competition were domestic as well as foreign designs. The benchmark for the testing would be the Thompson 1928A1. Also, the German MP35 (Bergmann) and MP40, as well as the British Sten and Finnish Suomi M/31, participated in the testing, but for research purposes only.

HYDE MODEL 109

The "model 109," in .45 caliber, was the first of many submachine guns that George J. Hyde designed over the years. It was also the first to be tested in the trials at Aberdeen (October 5–27 and November 21, 1939). With its front pistol grip, cooling ribs, and compensator, it resembled the Thompson M1928 submachine gun. The cocking handle, however, was not located at the side or to the top of the receiver but protruded from the end cap. With each shot, it moved ½ inch toward the eye of the shooter. Although the testers saw no immediate danger of injury, the feature presented a mental hazard. In the mud and dust tests, the model 109 proved superior to the Thompson. Its recoil was less, and the gun cooled down more quickly after sustained fire. However, since the trigger mechanism caused problems and the bolt broke, and considering the unusual cocking handle, the weapon was not tested any further.

SMITH & WESSON LIGHT RIFLE

At the beginning of the tests, a semiautomatic rifle in 9 mm caliber was entered by Smith & Wesson. Chief engineer Edward Pomeroy and his team had developed this straight-blowback-operated firearm. The company wanted to find out how their weapon would perform from the perspective of the Army, and whether the development should be continued to create a full-automatic submachine gun in .45 caliber. The only remarkable feature was an extended ejection port cover that caught the empty cases and also housed a detachable box magazine. The magazine was inserted from the front instead of the bottom. The weapon was very well made, making the manufacturing process time consuming and expensive. After testing on November 28, 1939, the weapon was returned along with a list of requests and suggestions for improvement. Smith & Wesson did not pursue the matter, and no improved submachine gun was submitted for further testing.

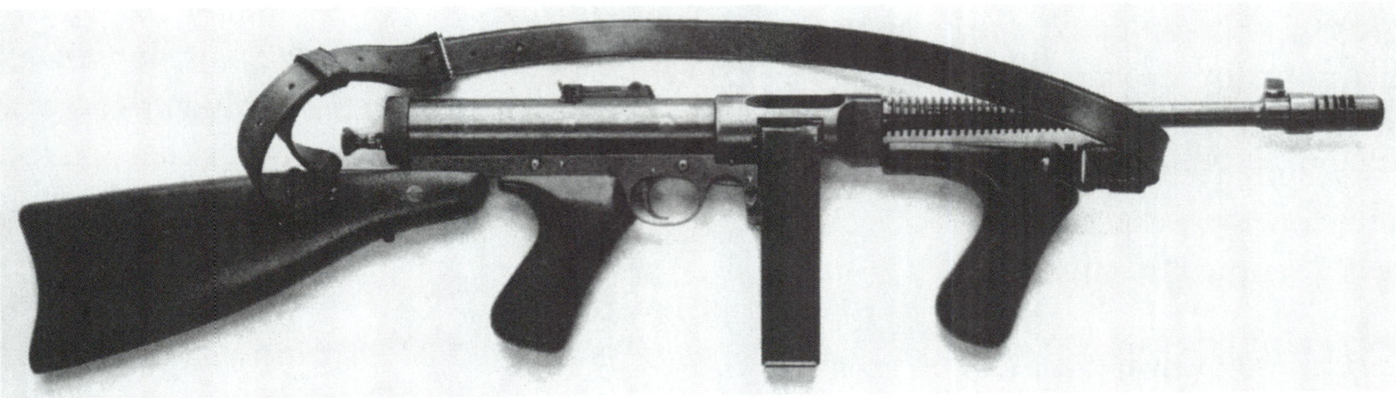

The model 109 designed by George Hyde. *APG*

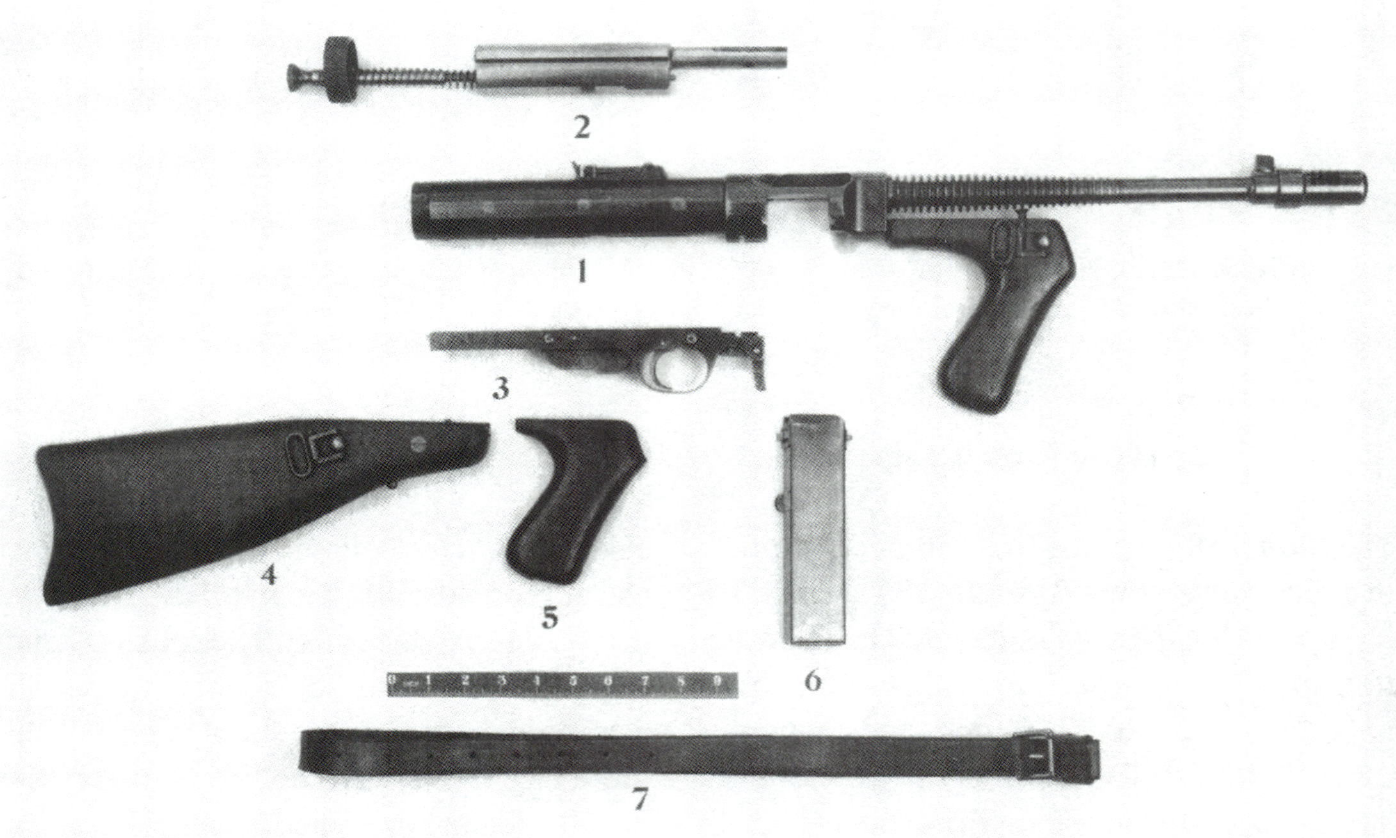

The cocking handle was not fitted to the side or top of the receiver, but to its rear end. With every shot it moved 1/2 inch toward the eye of the shooter. *APG*

Smith & Wesson light rifle. The spent cases were caught in the large magazine well.

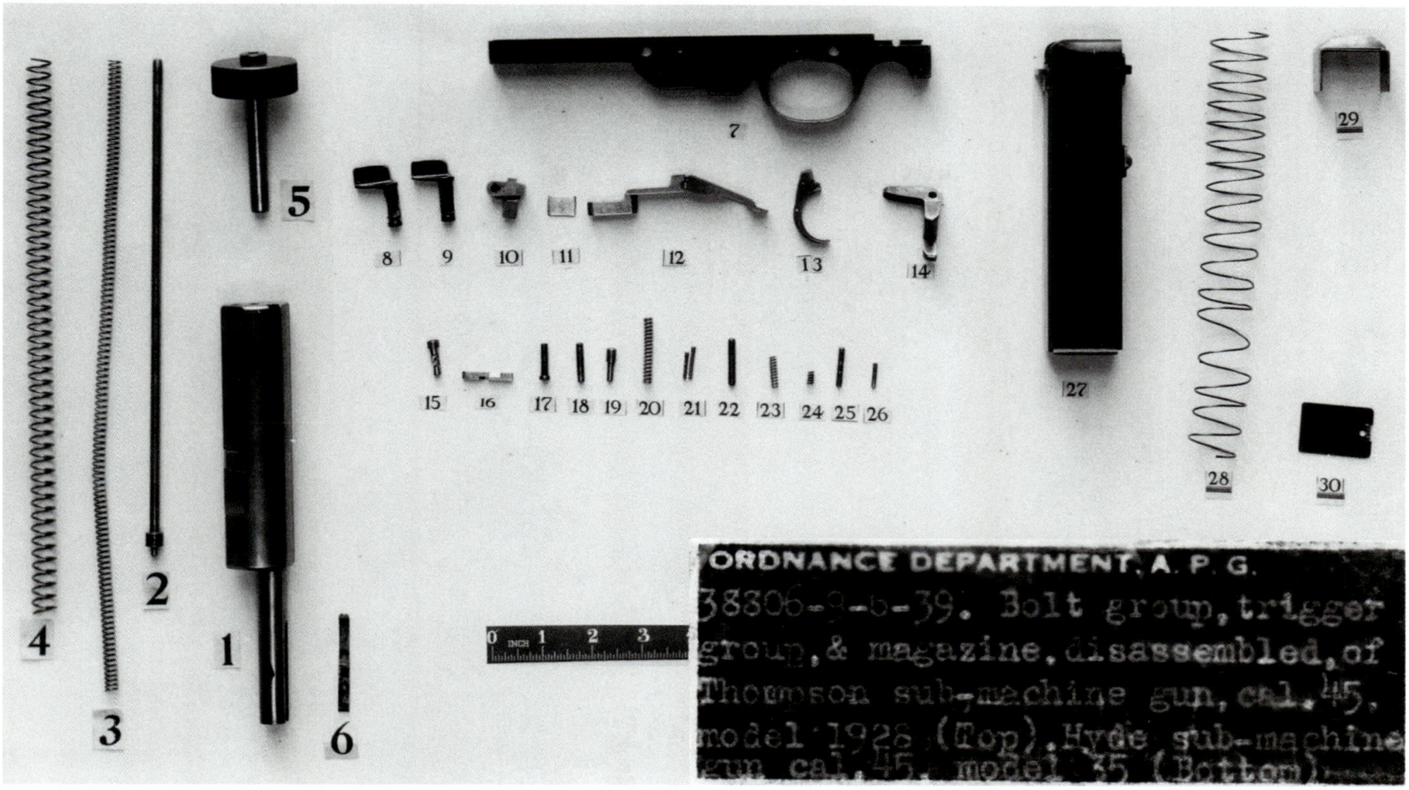

Disassembled bolt, trigger group, and magazine of the model 109. *APG*

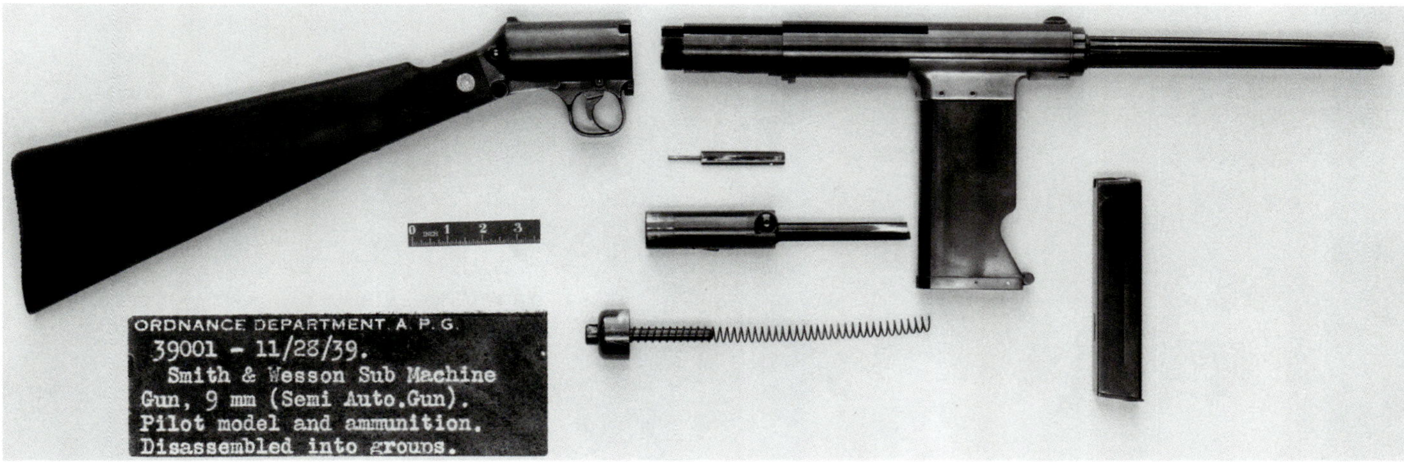

The model Mark I field-stripped, November 28, 1939. *APG*

REISING

The "H&R Reising," in .45 caliber, developed by Eugene Reising and manufactured by the Harrington & Richardson Arms Company in Worcester, Massachusetts, was tested on July 14–15, 1941, at Aberdeen. The weapon performed better than the Thompson, and in 1942 it was adopted by the US Marines as the Reising M50 (wooden stock) and M55 (with folding stock and without a muzzle brake). In combat, the sensitivity of the weapon against dirt caused significant problems. Why this was not further investigated during the tests at Aberdeen will probably remain a mystery. However, the Reising was not tested any further at Aberdeen, and an improved model was rejected after only one day of testing (September 13, 1943).

SUOMI M/31

The first foreign weapon in this testing phase was the Finnish Suomi M/31 (Suomi Konepistooli KP/31), in 9 mm caliber. It was tested from July 18 to 26 and again on August 2, 1940. The Suomi was highly praised for handling, accuracy, and manufacturing quality. Only the muzzle rise in full-auto mode and the awkwardly placed fire mode selector attracted criticism. A direct comparison with the Thompson was not possible because of the different calibers.

Position when shooting the S&W light rifle. The magazine well was used as a fore grip.

Picture from the manual showing how to carry the weapon for quick use

Comparative test of Reising M55 and Thompson 1928A1 at Quantico. The Reising performed better in the tests but was too sensitive to dirt in combat. *Courtesy of the National Museum of the Marine Corps, Quantico*

Both Reising models compared: the M55 (*on the left*) and M50 (*on the right*). *Courtesy of the National Museum of the Marine Corps, Quantico*

The Finnish "9,00 konepistooli M/31," also called the "Suomi." This 9 mm weapon was tested only for comparative purposes. *Courtesy of Leszek Erenfeicht*

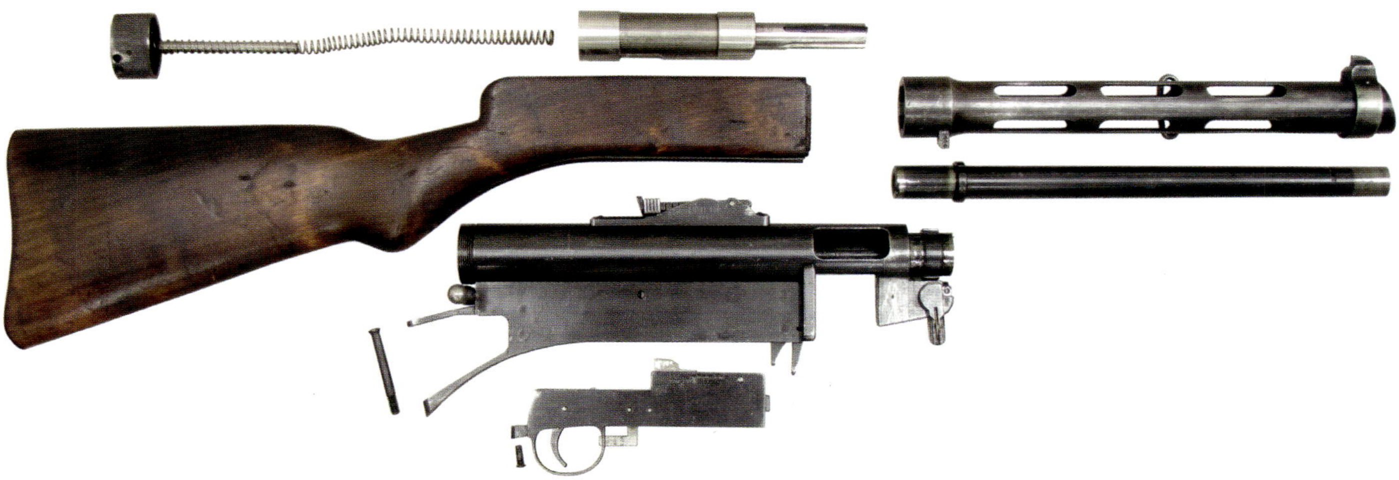

After the barrel jacket of the M/31 is unlocked, the barrel can be removed without the use of tools. *Courtesy of Leszek Erenfeicht*

STAR

The "Star," in 9 mm caliber, was a design of the Spanish company Star Bonifacio Echeverria S.A., from Eibar. It was manufactured in Spain and sold through the Atlantic Importing Company in the United States. That way, Aberdeen received an example, which was then tested from August 23 to 24, 1940. The designer, as written in the official paperwork, was José Cunill de Figuerola, but in fact, he was only the sales agent of the Atlantic Importing Company. The weapon shot from the open bolt with a delayed-blowback operation. The most interesting feature was the possibility to adjust the cyclic rate. Two levers were located on the left side of the receiver. The forward lever could be positioned for single-shot or low-speed full-automatic (300 rounds per minute) operation, whereas the rear lever had three positions: semiautomatic fire, high-speed full-automatic fire (700 rounds per minute), and safe. In the tests with dust and mud, the Star performed reasonably well. But it totally failed in the accuracy test. Of the thirty rounds fired, only seven hit the 6-by-6-foot target. Moreover, since the weapon did not show many advantages over the Thompson, except for its lower weight and the somewhat reduced recoil, it was not recommended for further testing.

ATMED

The "Atmed," in .45 caliber, was another development by George J. Hyde. It was manufactured by the Atmed Manufacturing Company for the competition at Aberdeen. The weapon took part from December 3 to 7, 1940. At first glance, it resembled the Thompson, but its weight was only 9.48 pounds and it operated on the common straight-blowback principle. The cartridge was ignited before the bolt completely reached its foremost position. That way, the recoil was reduced. The disadvantage, however, was the loss of gas pressure if the round was ignited too early, so that the bolt would not move back far enough to feed a new cartridge from the magazine. No compensator or muzzle break was attached. Nevertheless, the weapon could be fired with one hand in sustained fire, without noticeable recoil and without muzzle rise. The Atmed was the first submachine gun that passed the mud test successfully. But problems with double or even triple shots occurred in semiautomatic operation. Various alterations of the mechanics brought only little success, and the problem could never be totally eliminated during the tests.

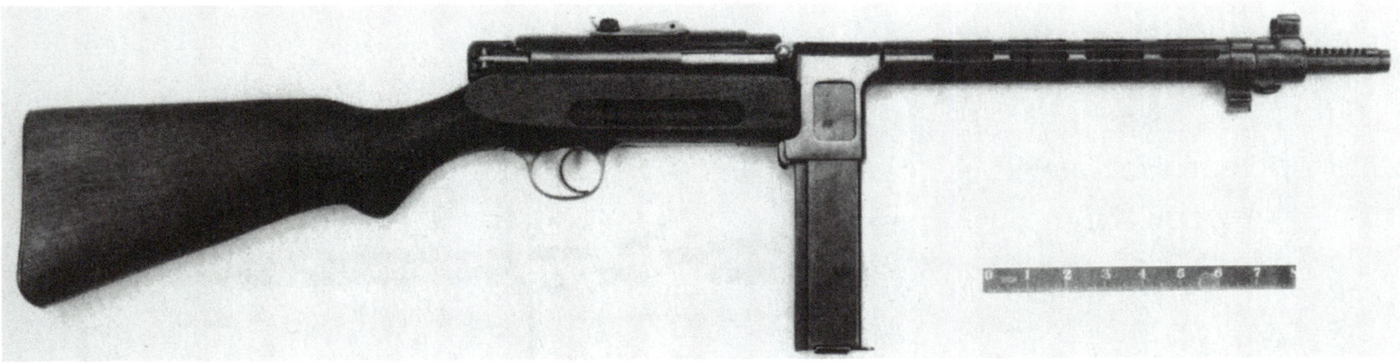

The "Star" of the Spanish company "Star Bonifacio Echeverria" in 9 × 19 mm caliber. *APG*

George Hyde's "Atmed" field-stripped. It was the first weapon to successfully pass the mud test at Aberdeen. *APG*

HI-STANDARD

The High Standard Manufacturing Company of New Haven, Connecticut, submitted a .45-caliber submachine gun named "Hi-Standard" to Aberdeen. It was tested on August 28–29, 1940. The weapon performed satisfactorily, but the rate of fire (900 rounds per minute) was too high in consideration of the twenty-round magazine. Further tests were recommended, and the weapon was returned to High Standard for revision. An improved second model was then tested from November 24 to 26, 1941. A new forty-round magazine came with the weapon, but unfortunately, the company had made a mistake in the design: the magazine could not be used because the forestock had been moved about 1.5 inches to the rear and interfered with the insertion of the new magazine. The rate of fire remained almost unchanged, with 800 rounds per minute. Although the Hi-Standard passed the dust test without problems, it failed in the mud test. Overall, the weapon was considered useful, but further testing was necessary. From 1942 onward, the company Marlin Arms in New Haven, Connecticut, manufactured this weapon as the "United Defense UD M42." However, disputes among the three companies, High Standard, United Defense, and Marlin Arms, disgruntled the War Department and prevented a greater commercial success.

Two Slovenian partisans posing with their UD M42 for a propaganda picture

The Hi-Standard field-stripped. This weapon was the precursor of the United Defense M42. *APG*

The "United Defense UD M42" with its characteristic double magazines. *Courtesy of the Rock Island Auction Company*

WOODHULL

The Woodhull Corporation submitted a weapon that was based on the Winchester Model 1910 self-loading rifle. The modified design was converted to .45 caliber, fed from a ten-round magazine. But the whole design was extremely complicated, difficult to disassemble and reassemble, and costly to manufacture. The fire selector lever could be operated only when the bolt was retracted. The weapon was tested on August 15–16, 1942, and performed quite well. All failures were blamed on the magazine, which was very sensitive to dirt. Eventually, the weapon was rejected because of its complicated design.

The weapon of the Woodhull Corporation in .45 caliber. *APG*

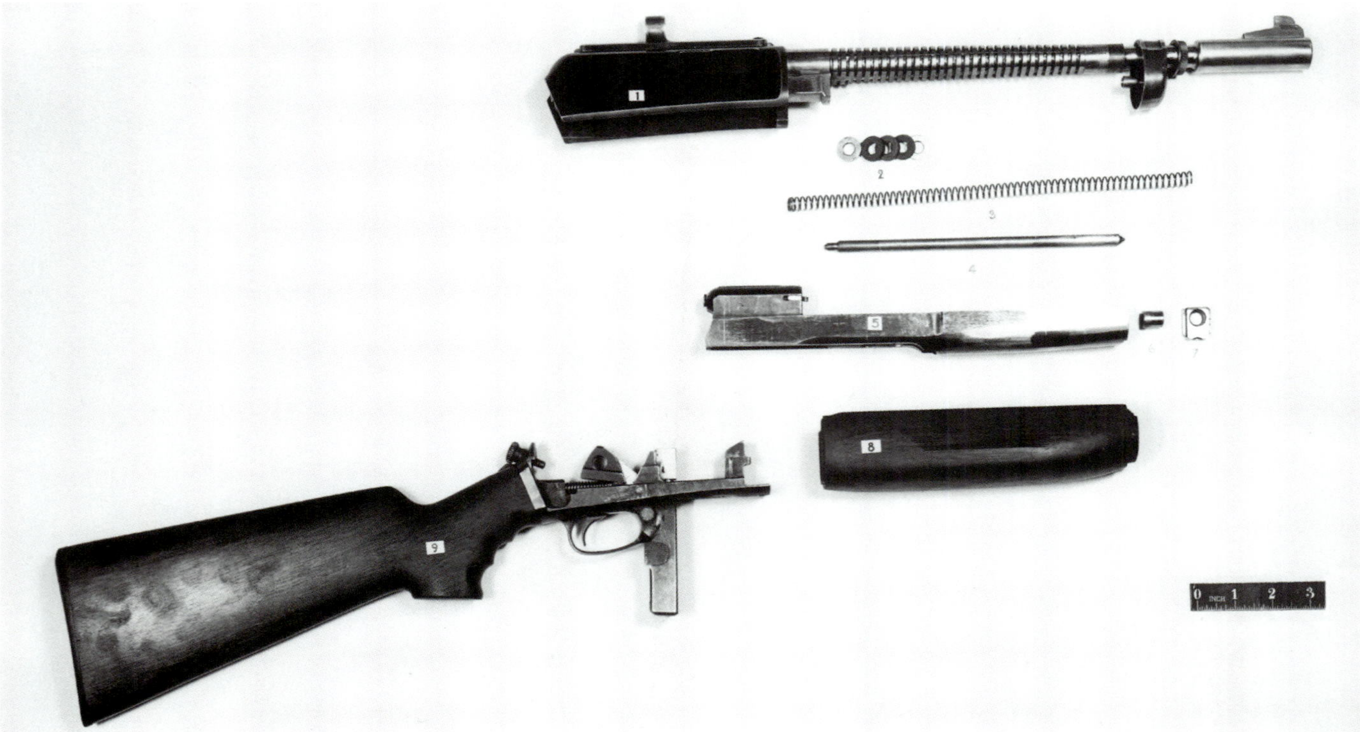

The Woodhull was dropped from the competition because of its complicated design. *APG*

STEN MK. II

The British Sten Mk. II submachine gun (9 mm caliber) was tested on January 23, 1942, for comparison purposes. The Sten was very insensitive to dirt. However, the testers did not like the magazine, which was inserted from the side, and the unusual position of both hands when shooting. In addition, a reliable safety mechanism was missing to prevent accidental firing. The extremely simple design of the Sten, with all its sheet-metal parts, had a significant impact on the development of the M3.

On February 14, 1942, Aberdeen Proving Ground published a report on the previously conducted tests. Each weapon was rated on a point system, with a maximum of 100 possible points:

- 15 possible points for weight
- 15 possible points for simple and economical production methods
- 30 possible points for reliability
- 10 possible points for accuracy in semiautomatic shooting
- 30 possible points for precision in full-automatic shooting

In general, the weapons in 9 mm caliber performed better in terms of accuracy and reliability than those in .45 caliber. Therefore, the report stated that "9 mm is not standard in the US Army and that 9 mm weapons should not be considered unless essential in obtaining a satisfactory submachine gun, which is not the case."

At the conclusion of testing, the .45-caliber weapons Atmed, Hyde Model 109, and Hi-Standard scored the most points. However, all the weapons still showed serious deficiencies, such that none of them could be considered as acceptable. The Atmed was highly praised and a more intense test was suggested, but the problems with the semiautomatic operation could never be eliminated.

Nevertheless, the testers were satisfied with the previous testing. They had gained a lot of experience, were able to compare different operating principles, and got ideas for future developments (muzzle brakes, buttstocks in direct extension of the bore, and so on).

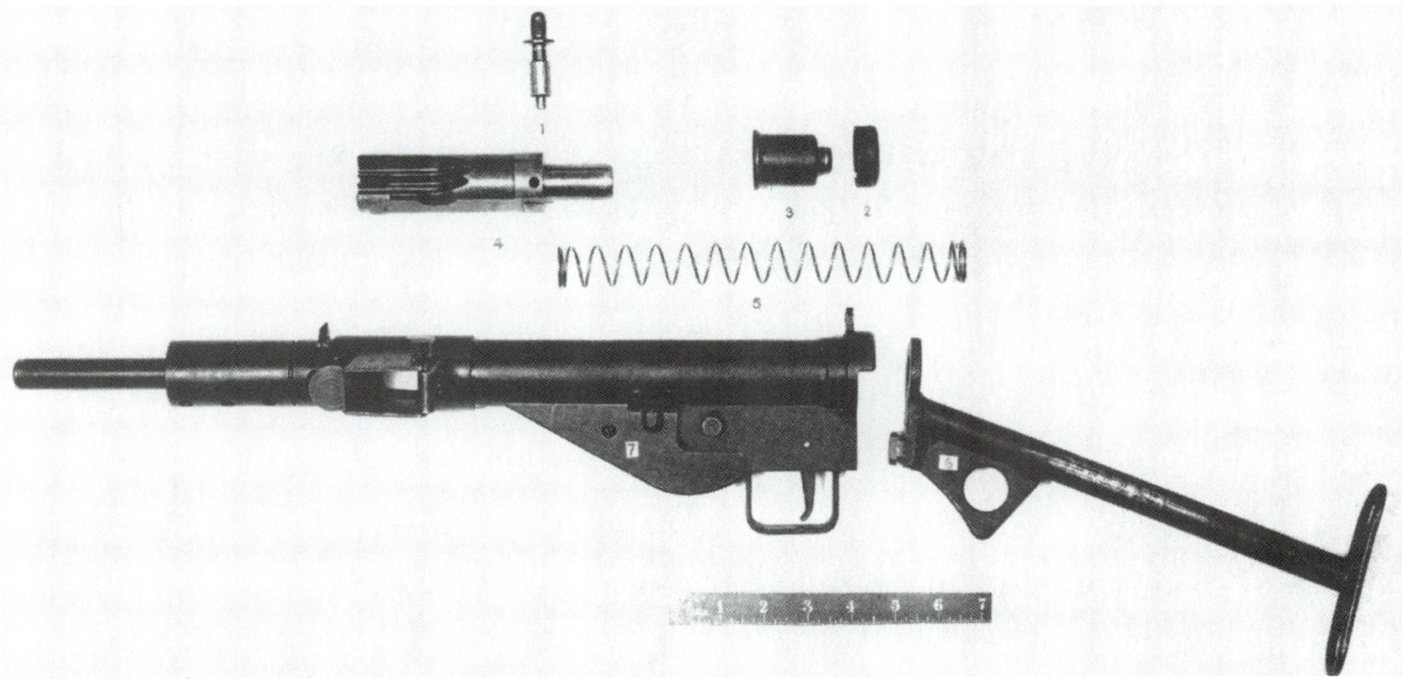

Also a Sten Mk. II (9 mm caliber) was tested for comparative purposes. *APG*

The simple design of the Sten gave important suggestions for the later development of the M3.

Some interesting ideas, such as this buttstock in direct extension of the bore, were also tested at Aberdeen. *APG*

THE SECOND PHASE OF THE COMPETITION, 1942–43

Due to the many new insights that were provided, the Ordnance Department decided to take a more active role in the development of small arms, not only selecting from among the different models that were presented by the manufacturers. In cooperation with the Inland Division of General Motors, a new weapon was developed from scratch. It took part in the competition at Aberdeen in November 1942, as the "T-20." The weapon and its development are described in detail later in this book.

Meanwhile, however, other manufacturers had submitted new models or revised and improved their already tested models:

HYDE-INLAND I

Following the rejection of the Hyde Model 109 in the 1939 tests, the weapon was thoroughly revised by the manufacturer. As the "Hyde-Inland I," it was tested at Aberdeen on April 2, 9, 10, and 15, 1942. The system was now better protected against the ingress of dirt. The barrel cooling ribs had been eliminated, and the cocking handle was relocated to the right side of the receiver. More than 6,000 rounds were fired, with only twenty minor malfunctions. Most of them were caused by the bolt hold-open feature, which did not work well: the bolt failed to remain in the open position after the last round was fired. Because of the weapon's satisfactory performance, the test personnel hoped to try out a revised, slightly improved model (see "Hyde-Inland 2").

MASCO

In April 1942, Aberdeen received a submachine gun delivered by the Masco Ordnance Company. It was a blowback-operated .45-caliber weapon, which had not yet been fully developed. The fire selector lever to switch to semiautomatic operation was blocked, and the weapon was missing any sights. After attaching a provisional sight, 200 rounds were fired before the test was canceled and the weapon was sent back to the manufacturer. The report stated that the weapon was "entirely unfit for military use."

MODIFIED PISTOL 1911A1

A rather unusual contribution to the tests was submitted by the Colt's Patent Firearms Company: a modified 1911A1 automatic pistol converted to fire full automatic. The weapon featured a 9.5-inch barrel with a perforated barrel jacket. An extended twenty-round magazine was provided, and a similar spare magazine could be carried in the wooden buttstock. Information on the test results are not available. But on May 28, 1942, it was decided that the development of this 1911A1 should no longer be continued.

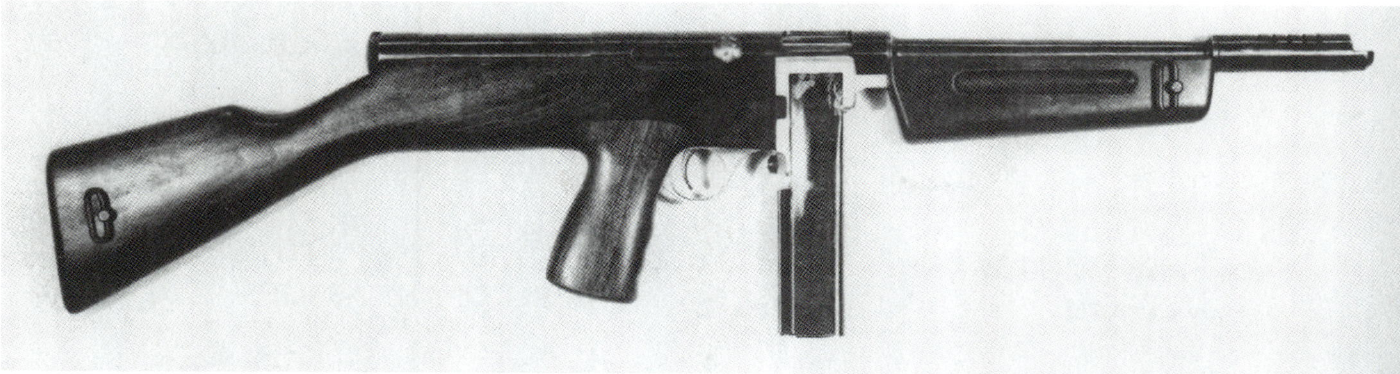

The weapon of the company Masco was dropped as "absolutely unusable" from the competition. *APG*

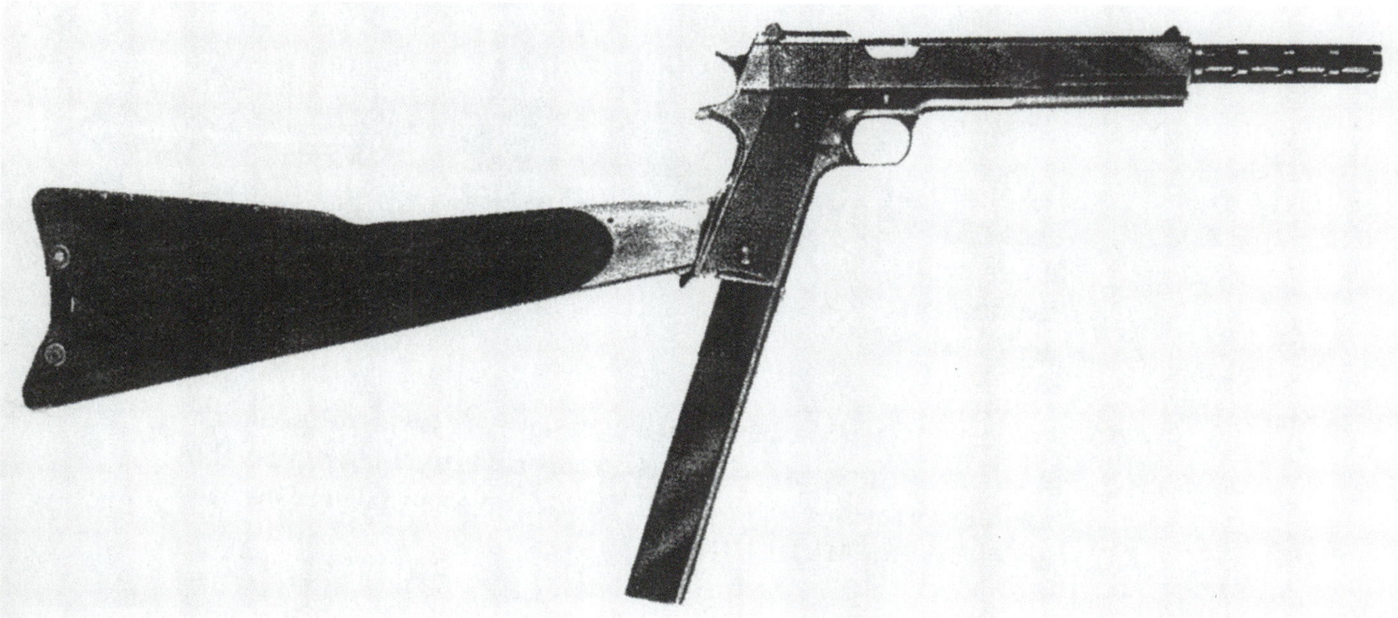

A very unusual competitor was the modified automatic pistol 1911A1, with extended barrel and perforated barrel jacket. *APG*

HYDE-INLAND 2 / HYDE-INLAND (M2)

A revised version of the Hyde-Inland I was tested from June 19 to 22, 1942; five weapons had been submitted to Aberdeen. It was recommended that the magazine well be deepened so as to accommodate the larger thirty-round magazine, and that a stronger bolt-driving spring should be used. Since the weapon was otherwise rated as very good, the Ordnance Department recommended its adoption as the new "U.S. Submachine Gun M2." However, as mentioned previously, difficulties in the manufacturing process thwarted this plan.

STEN MK. III

Since the first tests of the British Sten Mk. II aroused strong interest in its simple design, the successor—the Mk. III—was also sent to Aberdeen for comparison purposes. From August 22 to 25, 1942, the weapon was tested, but due to the lack of 9 mm ammunition, only 932 rounds could be fired. The Sten passed all the tests without problems, the recoil was very soft, and the accuracy was good. But again, the sideways-inserted magazine was criticized because it made handling and shooting difficult. A safety mechanism was missing to prevent the weapon from discharging if it was accidentally dropped to the ground. But the Ordnance Department was made aware of how easily and cheaply a useful submachine gun could be manufactured.

TURNER

The submachine gun designed by J. Russell Turner of Butler, Pennsylvania, was the only gas-operated weapon that was tested as part of the competition at Aberdeen. The gas required to operate the weapon was bled off a port on the left side of the barrel, a short distance ahead of the chamber. The Ordnance Department was curious to see how such a functional principle would work in a submachine gun. So, the "Turner" (.45 caliber) was tested from August 26 to 28, 1942. The first problem that occurred was the enormous heat: several of the testers burned their fingers on the hot cocking handle and the rivets that secured the forestock band. A total of 5,500 rounds were fired with only some minor problems, mainly caused by the troublesome magazine. These were in fact adapted Reising magazines. The rate of fire was extremely high, with approximately 1,400 rounds per minute. Although the "Turner" was inferior to the Thompson and the Hyde-Inland M2, it was rejected because of its complicated gas-operated design.

The Hyde-Inland M2 was proposed by the Ordnance Department as the new "U.S. Submachine Gun M2," but production difficulties prevented the adoption. *APG*

The Sten Mk. III performed well on the shooting range, but testers did not like the magazine inserted from the side, and the position of the hands when shooting.

The gas-operated submachine gun designed by Russell Turner. It was too complicated for military use. *APG*

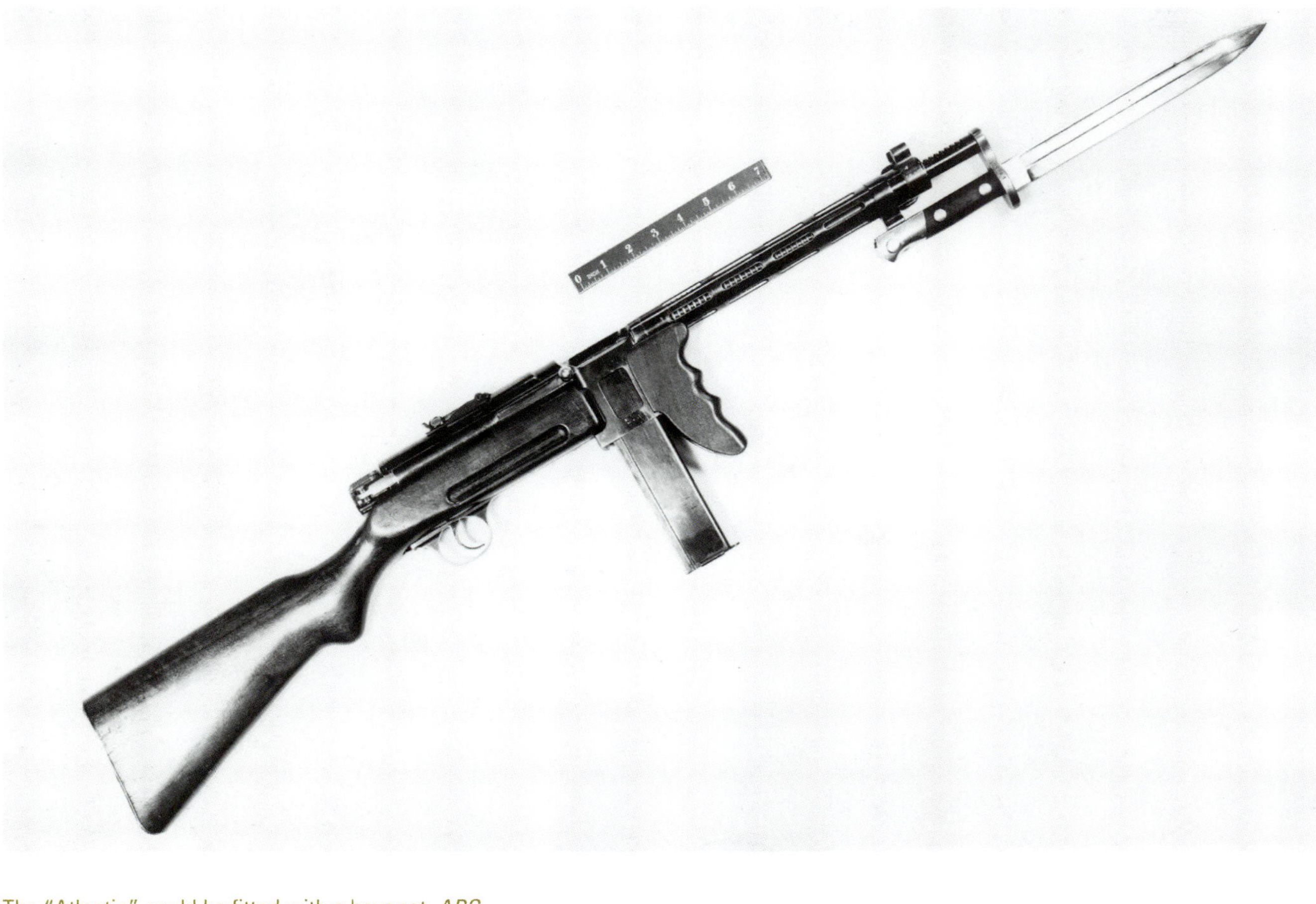

The "Atlantic" could be fitted with a bayonet. *APG*

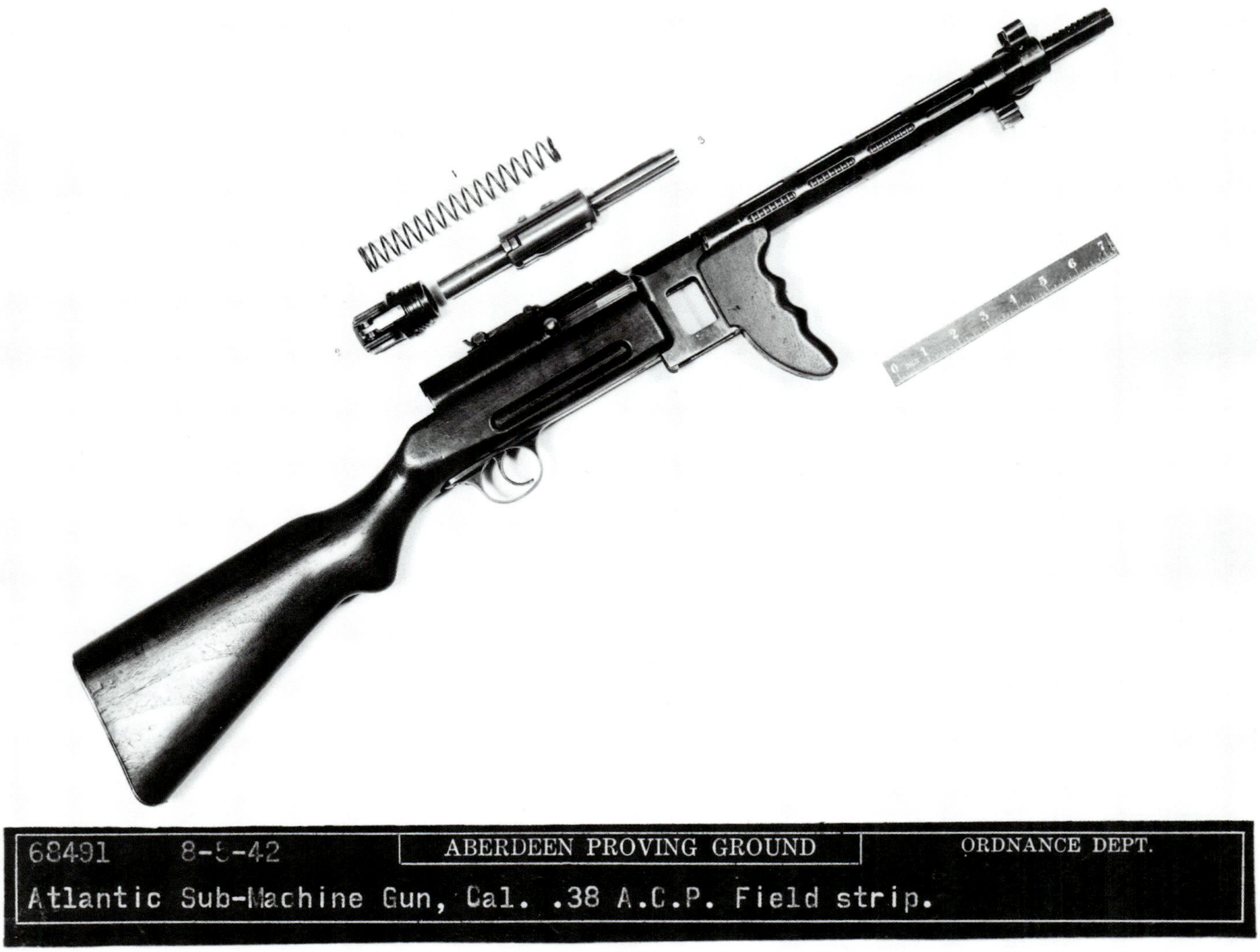

The "Atlantic" was the second attempt of the company Star Bonifacio Echeverria, this time in the unusual .38 ACP caliber. *APG*

ATLANTIC

Two years after the "Star," the company Star Bonifacio Echeverria S.A. submitted another submachine gun to Aberdeen via the Atlantic Importing Company, this time in the unusual .38-caliber ACP, which was not a US Army standard cartridge. The "Atlantic" was a delayed-blowback-operated weapon with a fire selector switch for semiautomatic operation. The rate of fire was a little more than 500 rounds per minute; the magazine held thirty rounds. The weapon was tested on August 28–29, 1942, scoring seventy-seven hits out of one hundred shots on the target. The recoil was considered acceptable. However, the Atlantic showed no advantages that would have justified its adoption.

AUTO-ORDNANCE T-2

Concerned about losing the lucrative government contract over the Thompson submachine guns, the Auto-Ordnance Corporation submitted its own candidate for the Aberdeen tests. The newly developed "T-2," in .45 caliber, fired from the open bolt. It was primarily made from simple sheet-metal stampings and was embedded in a wooden stock, which reached almost to the muzzle. The test began on September 9, 1942, under the watchful eyes of Capt. Peter White (Ordnance Department) and John Ferguson and James Landing (both from Auto-Ordnance). Since at that time the Hyde-Inland (M2) was already favored, the T-2 had to compete against this weapon in direct comparison—and it performed poorer in most tests.

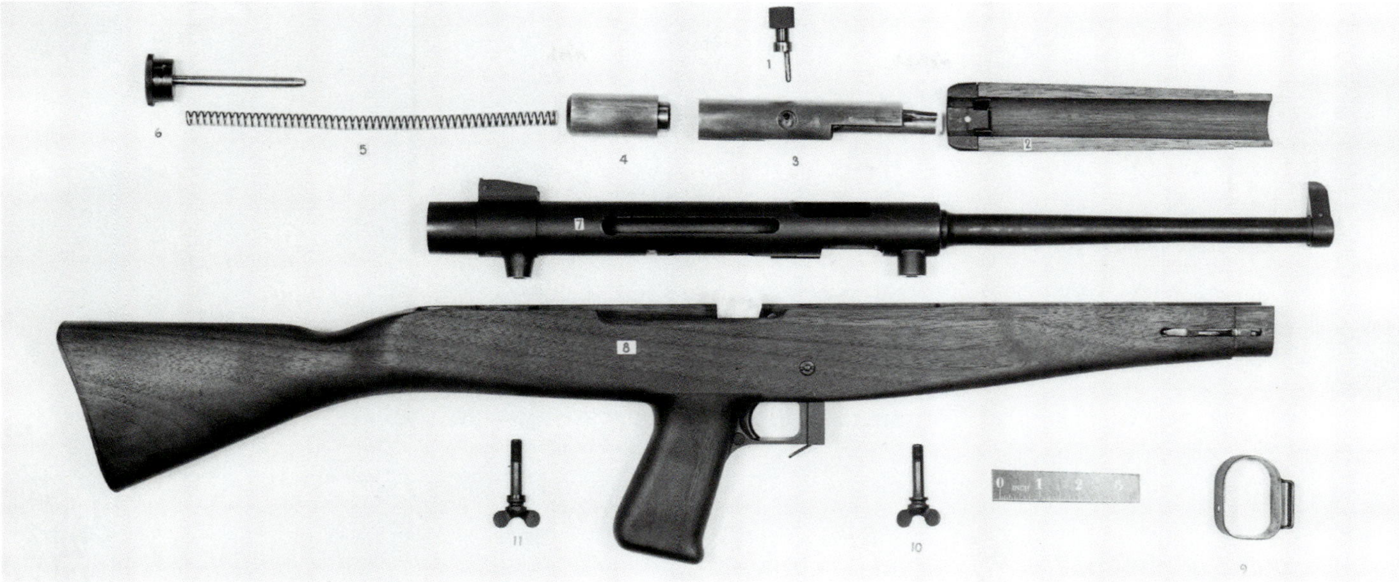

Auto-Ordnance hoped to continue the success of the Thompson models with their new T-2. *APG*

The Australian "Austen" (9 mm caliber), based on the Sten design, was tested for comparative purposes. *MOD*

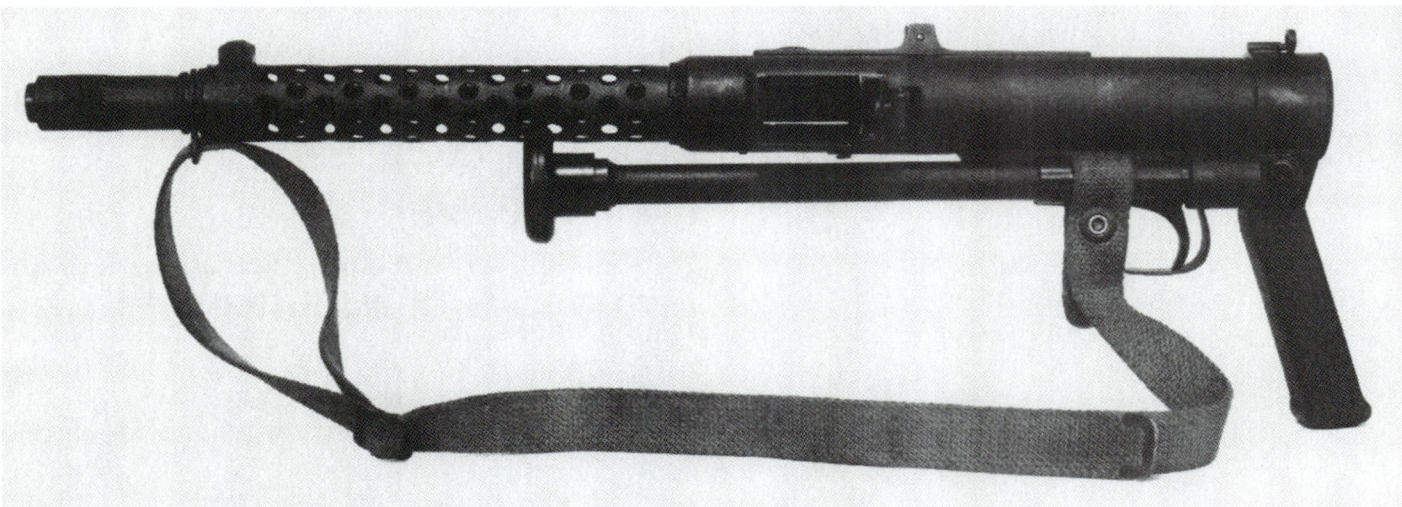

The "UD-1" could be adapted for the 9 mm or .45 calibers with only a few simple steps. *APG*

The respective barrel that was not in use could be screwed in a mount under the receiver. *APG*

AUSTEN MK. I

The "Austen" was (as the name suggests) an Australian development based on the British Sten in 9 mm caliber. Both the recoil spring, located in a three-part telescopic tube with firing pin, and the folding stock were copied from the German MP40. The weapon was tested from November 25 to 27, 1942. There were hardly any complaints; only dirty magazines caused some feeding problems.

UNITED DEFENSE UD-1

An interesting weapon was submitted by the United Defense Corporation. The "UD-1" could be adapted for the 9 mm or .45 calibers with only a few simple steps. This was accomplished by changing just the barrel; the same bolt was used. The respective barrel that was not in use could be screwed in a pivoted mount under the receiver. When pivoted to the rear, this barrel served as a simple shoulder stock. The weapon was tested on April 2, 1943, with only 180 rounds in .45 caliber fired (rate of fire 1,150 rounds per minute). No tests were done with 9 mm ammunition. The accuracy was very poor, and the weapon did not pass the mud test in two attempts.

OWEN

The "Owen" was the standard submachine gun of the Australian army, made by the Lysaghts Newcastle Works. It was named after its designer, Lt. Evelyn Ernest Owen. The weapon was a straight-blowback select-fire design in 9 mm caliber. Its characteristic feature is the thirty-three-round stick magazine, inserted from above, and because of that, the sights have been placed offset to the right side. Empty cases were spent downward from the weapon. The Owen participated in the tests from May 1 to 4, 1943. The weapon shot very accurately, and the dust and mud tests caused no problems (except one feeding failure). In total, 200 rounds were fired without any complaints after the weapon was exposed to water spray simulating heavy rain.

OLSEN

The "Olsen," named after its designer, A. W. Olsen, was a straight-blowback-operated weapon in .45 caliber. Most interesting was the unusual design of its twenty-round magazine: a curved lip permitted the rounds to rise directly in the path of the bolt rather than to be picked up by the bolt on its forward stroke. The problem was that the magazine could not be released if the bolt was in its closed position, and in case of a malfunction, the weapon became useless. To fire in full automatic, the trigger and an auxiliary button on the left side of the stock had to be pressed simultaneously. The test of the weapon began on June 2, 1943, and ended on the same day. Only 240 rounds were fired; the recoil was characterized as "very severe." The trigger vibration caused pain in the operator's finger. The accuracy test was a disaster: only five shots out of twenty fired hit the paper target, with none striking the bull's-eye. Of the next twenty rounds fired, not a single bullet hit the target. After this result, the test of the Olsen was canceled.

While the tests were still not finished, Aberdeen Proving Ground on March 10, 1943, published the second report about the results that had been achieved so far. Again, the experience gained was praised, and it was noted that the submachine guns of the future would probably be made mostly of sheet metal. Of all the weapons that were tested, the T-20 performed best by far.

Perhaps everything would have been different if the US Army had not stuck to the .45 caliber so vehemently. In both competition phases, the weapons in 9 mm caliber dominated. They were light, precise, and technically simple designs. Their ammunition was the most widespread in the world for pistols and submachine guns. But the Ordnance Department stuck to the powerful .45 ACP cartridge, whose bullet supposedly had better stopping power due to its mass.

Comparison of a .45-caliber cartridge (*left*) and a 9 × 19 mm cartridge (*right*)

Evelyn Owen with some of his weapons. Picture taken ca. 1940.

Characteristic features of the Owen are the stick magazine inserted from above and the sights placed offset to the right side. Many guns were painted in green or camouflage colors. *Courtesy of the Rock Island Auction Company*

The Australian Army used the Owen in large numbers. Lysaght in Newcastle made over 45,000 pieces from March 1942 until September 1945.

Its severe recoil and great dispersion made the Olsen unsuitable for military use. *APG*

04 DEVELOPMENT OF THE T-15 AND T-20

THE FIRST TEST PHASE of the various submachine guns at Aberdeen was less satisfactory than expected. Most weapons showed more or less serious shortcomings, so that no really promising model could be found. In particular, the experiments with dirt and mud caused many weapons many problems—they were simply unsuitable for use in military scenarios. The high cyclic rate and the resulting muzzle climb, when fired in full-automatic mode, were criticized.

Nevertheless, the competition was a benefit for all participants because it brought many new insights, especially the simply manufactured "sheet-metal weapons" such as the British Sten, which left lasting impressions. The Ordnance Department decided to henceforth participate in the development of new weapons and not just select between the presented models of the manufacturers.

For this purpose, the Ordnance Department contacted the Inland Division of General Motors. Three individuals were responsible for the "T-15" project:

- Frederick W. Sampson, chief engineer of the Inland Division of General Motors
- George J. Hyde, head of engineering and holder of numerous patents
- Col. René Studler, US Army Ordnance Research & Development (R&D) officer

The original specification (OCM 19007) of the project, from October 8, 1942, was altered shortly after its completion, adding the demand for a conversion kit to 9 × 19 mm caliber. This was the standard caliber not only of the British submachine guns but also the German ones. The new weapon could, therefore, be easily supplied to the British. Consideration was given to support resistance groups in the European theater of war; for example, by airdrops of containers. The groups could then provide themselves with German booty ammunition.

Another change was the elimination of the semiautomatic function to simplify the design. The rate of fire was so low that an operator with a little practice could release single shots just by trigger manipulation. The complex mechanism proved to be unnecessary.

Following these changes, the project received the new designation "T-20."

Five prototypes of the "T-20" and five conversion kits for 9 mm caliber were manufactured by General Motors until the end of November 1942, for the upcoming tests. The guns in .45 caliber weighed 7.66 pounds, and, converted to 9 mm caliber, the weight increased slightly to 7.74 pounds. The cyclic rate was 400 rounds per minute, thus almost completely eliminating the muzzle climb.

The first standard test was a 50-yard shooting on square targets (6 × 6 feet). The T-20 scored ninety-seven hits out of a possible one hundred, which was the second-best result of all tested submachine guns. Only the Hyde-Inland M2 scored ninety-nine hits. In contrast to both weapons, the Thompson 1928A1 had scored only ninety-three hits.

The following test was a 100-yard shooting, for which the weapons were immovably fixed to a gun rest. The .45-caliber T-20 showed a maximum spread of 9.36 inches. When converted to 9 mm the spread was reduced to 7.06 inches. For comparison, the Thompson 1928A1 spread about 10.57 inches in the same test.

Col. René Studler, US Army Ordnance R&D officer. *APG*

An ordnance officer is testing the M3. The wire stock could cause an uncomfortable feeling when pressed against the cheek.

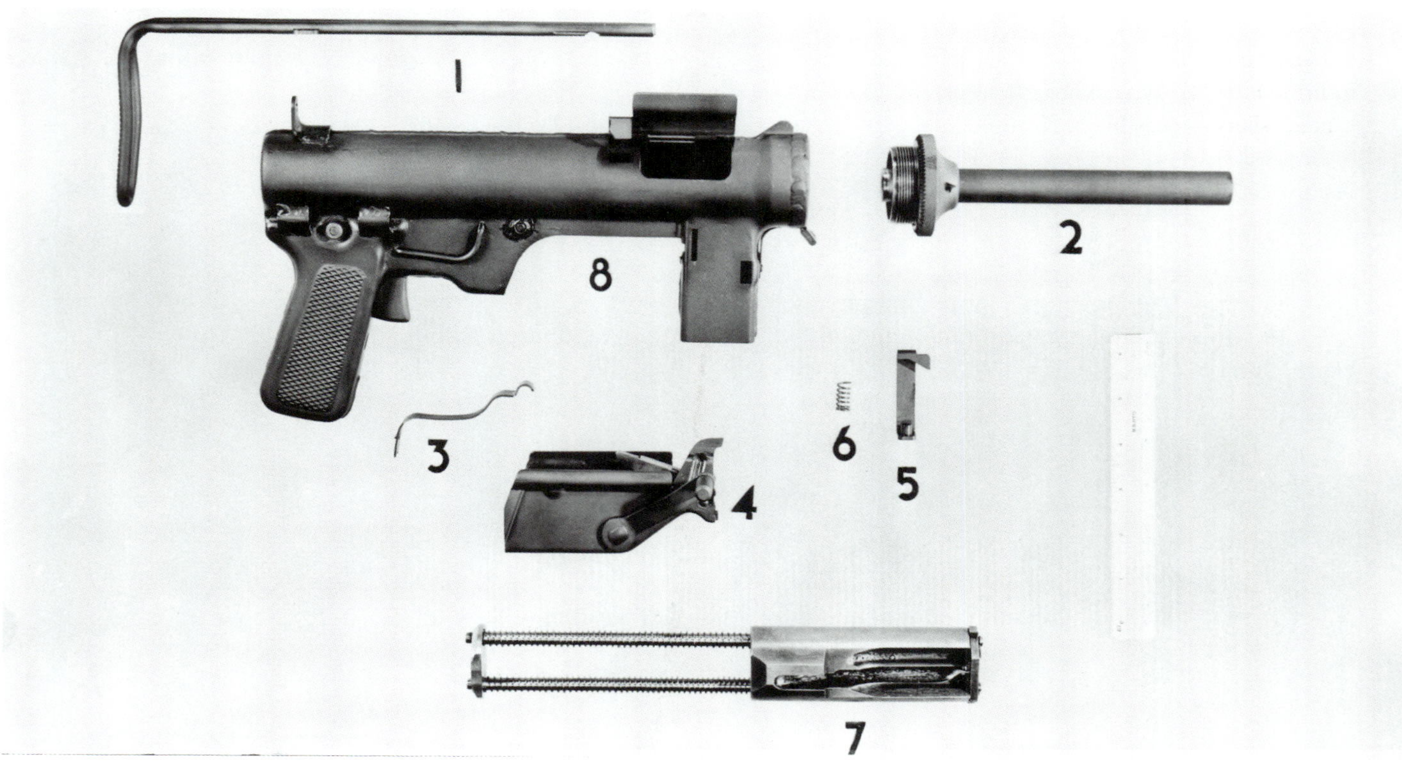

An early M3 taken apart (November 24, 1942). Note the early type of dustcover, of which a strip was cut and bent downward to serve as a "safety hook." Also the shape of the ejector housing differs from the later series production. *APG*

The next step was an endurance test: at first, 5,000 rounds of brass-cased .45 ammunition were fired. The weapons performed well, except for two failures caused by defective followers in the magazines. Thereafter, 500 rounds of steel-cased .45 ammunition were fired, resulting in three ejection failures and one feeding failure caused by the magazine.

After converting the weapons to 9 mm caliber, the tests were repeated with commercial ammunition made by Winchester. A total of twenty-three failures to eject occurred before the test had to be interrupted after 3,750 shots. It was determined that the ejection port was too small. The cases often hit the edges and fell back into the receiver, thus blocking the bolt. After the ejection port was enlarged and a slight alteration of the ejector's protrusion was made, the remaining 1,250 rounds of the test were fired without any problems.

In the final part of the test, the T-20 had to face dust and mud, and it proved to be the most reliable weapon of all submachine guns tested so far.

After the T-20 had performed so well in the tests at Aberdeen, some of the prototypes were distributed to military units for a series of intensive field tests:

- US Army Infantry Board, Fort Benning, Georgia
- US Army Armored Forces Board, Fort Knox, Kentucky
- US Army Airborne Command, Fort Bragg, North Carolina
- Amphibious Warfare Board, Fort Carrabelle, Florida

The infantry consumed 3,014 rounds in a series of seven different tests:

1. time required for disassembly and reassembly
2. accuracy: 10 targets, 10 shots per target
3. bursts: 3 to 5 rounds, 300 rounds per target
4. bursts: 5 to 10 rounds, 200 rounds per target
5. firing at 5 "E"-type silhouette targets in rotation as many shots as possible in 25 seconds
6. dust test
7. mud test

Firing the M3. Some spent cases are still in the air.

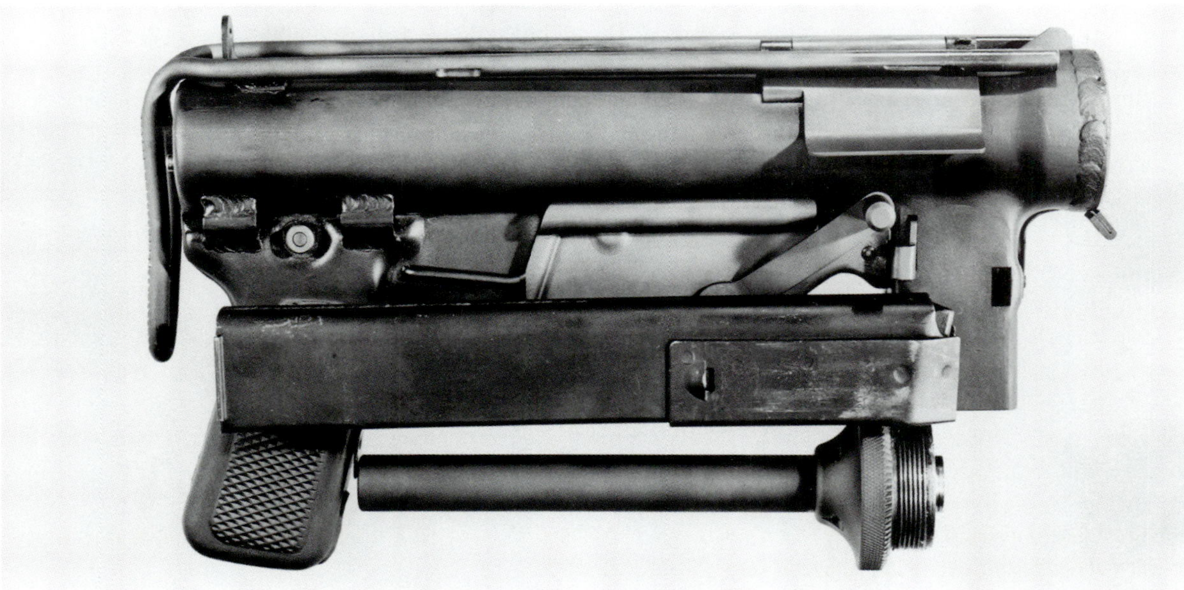

The disassembled early M3 ready for being packed. Note the early type of dustcover and the laterally welded mounts for the stock, which would later be replaced by one single piece in the series production. *APG*

This picture from February 6, 1943, shows the M3 in series design, except for the still-early type of ejector housing. The meaning of the marking "I.M.P" on the magazine is unknown. *APG*

The Infantry Board found the weapon to be reliable and accurate. Few ejection failures occurred when cases hit the edges of the ejection port, and two failures were caused by the cocking pawl obstructing the forward movement of the bolt. But there were major problems with the magazines: the follower was prone to jam in the magazine housing, causing feeding failures.

For the Armored Forces, the T-20 was an ideal defensive weapon for use in their vehicles. They preferred the handy weapon with its compact size over the bulky and heavy Thompson models. A total of 3,558 rounds were fired at Fort Knox, including a 2,000-round function test and dust and mud tests. Again, the weapon was praised, but the magazine was criticized.

The Airborne Command was also impressed by the small size and the low weight of the T-20. Twelve submachine guns could be stored in an "A-5 Aerial Delivery Container" for dropping by parachute. Regarding the shooting tests, the command was very economical—only 385 rounds were fired. And again, only failures caused by the magazines interrupted the tests.

The Amphibious Board was even more economical, with only 265 rounds fired. The weapon was fired at silhouette targets from a boat, then dropped into the water and fired again. Even after being buried in sand and dug up again, the weapon worked without any problems. The only failures were caused by sand jamming the follower of the magazines.

All participants reported back positively about the new weapon and recommended its adoption for military service. The outstanding point of criticism was the magazine: a rework of its design was highly suggested, but it never happened. The design remained unchanged until the end of production.

The adoption of the submachine gun was proposed in December 1942 and approved on January 11, 1943. That same month, a contract for 300,000 weapons was awarded to the Guide Lamp Division of General Motors, with the contract numbers "W-294-ORD-2107" and "W-33-ORD-825." And in May 1943, eight months after the project began, the series production of the now officially designated "U.S. Submachine Gun, Caliber .45, M3" started.

The preproduction models differ in some details from the later series weapons. On the rearward side of the dustcover, a strip was cut and bent downward. This "hook" served as a safety, blocking the movement of the bolt. In the series, the hook came as a separate part and was riveted to the inside of the dustcover. Also, the housing of the ejector and the crank mechanism was originally chamfered at its rear side and provided with a reinforcing bead.

STANDARDIZATION OF TESTS WITH DUST AND MUD

During the tests with various submachine guns at Aberdeen, there was no standardized method for carrying out the tests with dust and mud. Neither the amount nor the composition of the materials nor the duration of exposure was set uniform and binding. Although on the whole, the performance of weapons could be tested, the results were not fully comparable.

To the credit of Colonel René Studler, the chief of Ordnance Research & Development Center at Aberdeen, from October 1944 onward, weapons have been tested by standardized methods. Also, the already adopted weapons such as the Thompson M1928A1, the M3 "Grease Gun," and the M1 carbine underwent this procedure for comparative purposes.

THE DUST TEST

For this test, a 6-by-3-by-3-foot box, with a removable lid, was made from plywood. At one end, about 3 inches below the top, a hand-operated fan was installed (fan diameter of 12 inches). A 2-inch hole was drilled into the lid, through which the dust, sand, and other materials could be poured directly into the airstream for optimum distribution. The opening was located 15 inches forward of the fan nozzle. Two longitudinal struts on the inner sides were serving as a support for various crosspieces for positioning the weapon(s).

The material used was a mixture of 9 pounds of Albany sand (grade 0) with 1 pound of clean silica core sand. All (100 percent) of the mixture passed through a 30-mesh sieve, 80 percent through a 50-mesh sieve, and 3.4 percent through a 100-mesh sieve.

In preparation for the test, the weapon was cleaned and lightly oiled. The muzzle was sealed by tape. Weapons firing from the open bolt had a loaded magazine inserted, and the bolt was placed in the firing (rear) position. Closed-bolt weapons also had a magazine inserted, and, in addition, a round was chambered. The bolt was placed in the firing (forward) position. If the weapon was fitted with a dustcover, it was closed for the test.

The weapon was placed in the center of the box, with the stock showing toward the fan. Now the dust mixture was poured into the box at a rate of 5 pounds per minute, while the fan handle was cranked at a speed of sixty revolutions per minute. The weapon was exposed to the blowing dust in its topside-up position for one minute, then it was turned upside down for another minute. After the weapon was removed from the box, it was roughly cleaned by wiping by hand and blowing on the weapon. Then the weapon was fired with the dusty magazine inserted, and then again using a clean magazine and ammunition.

A wooden box that was used for dust tests in 1940, long before standardized methods were set up. The dust mixture was poured through the hole into the box while the fan handle was cranked by hand. *APG*

In October 1944, uniform rules were applied for the dust and mud tests. Here an M2 carbine and an M3 are waiting to get dusted. *APG*

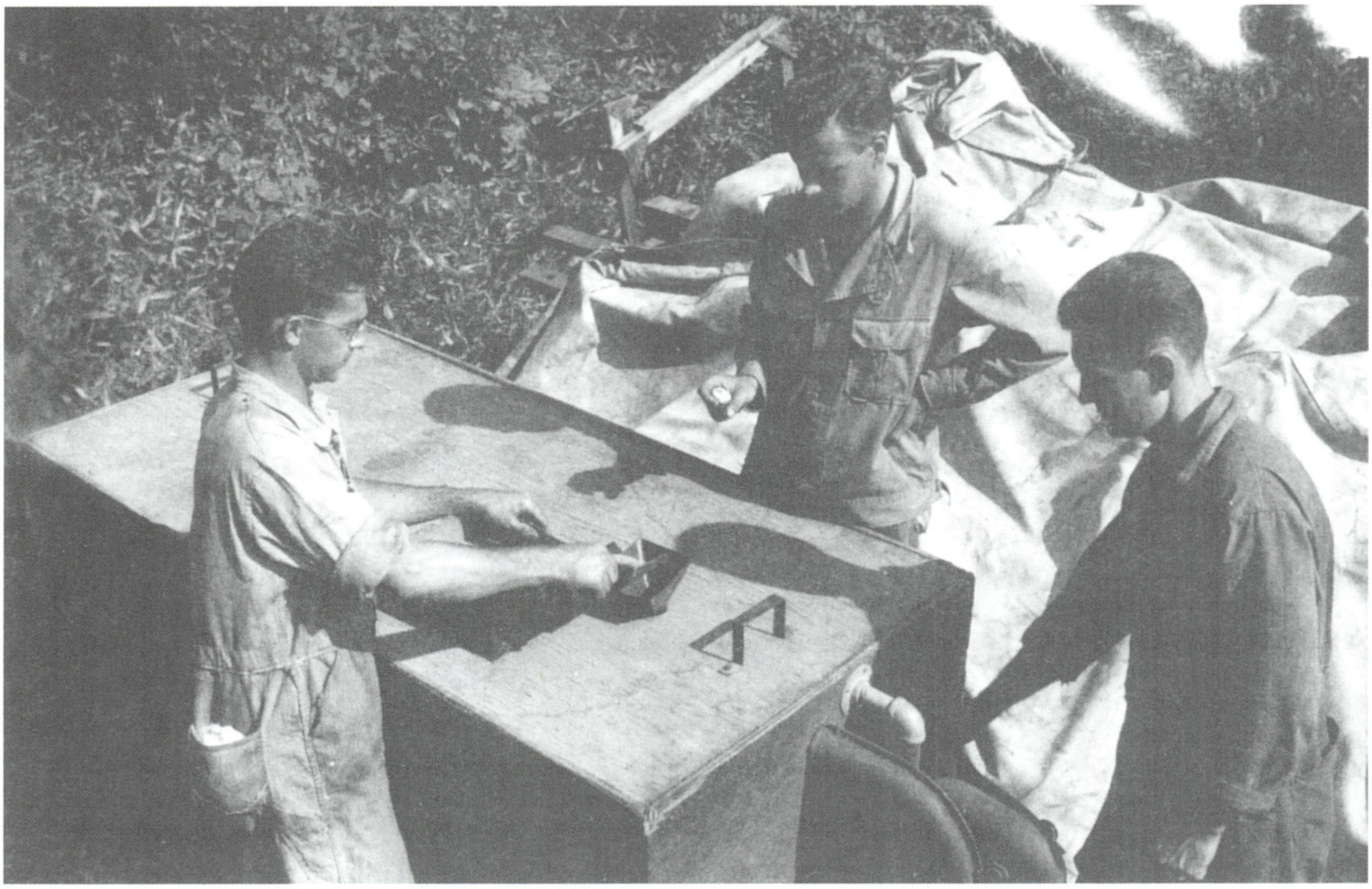

A worker is pouring dust into the funnel while a colleague is cranking the fan. Another colleague is keeping a watchful eye on the clock.

The M3 after being removed from the dust box. *APG*

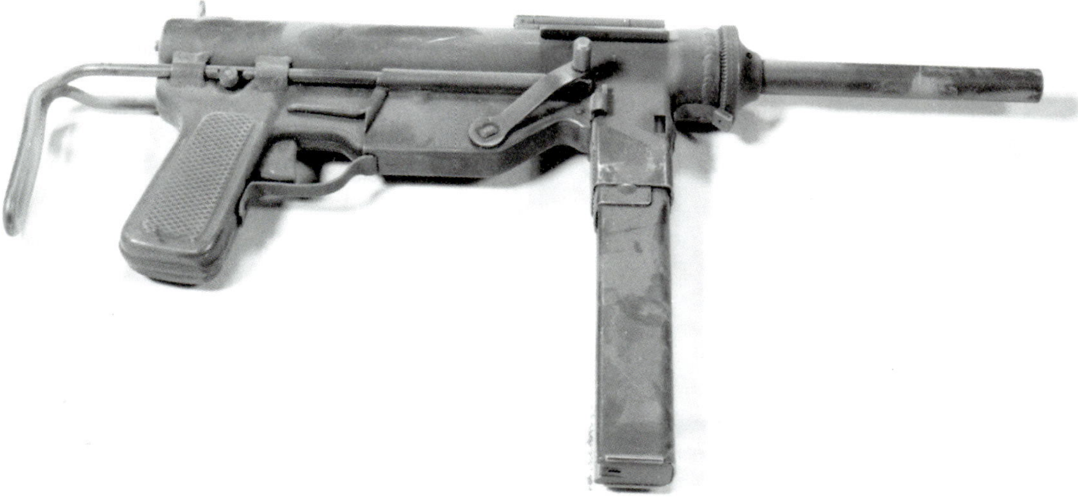

The Thompson was also exposed to dust for comparison purposes. *APG*

THE MUD TEST

The mixture used for this test consisted of 10 pounds of red clay (dried and pulverized), 2 pounds of clean river sand, and 8 quarts of water. The weapon, with inserted magazine, was dipped into the mud for fifteen seconds. After the weapon was removed from the mud, it was cleaned by simply wiping by hand and blowing on the weapon. Then the weapon was fired with the muddied magazine inserted, and then again using a clean magazine and ammunition.

The box for the mud test. *APG*

PATENTING

On May 1, 1944, Frederick Sampson and George Hyde filed three patent applications for the design of their submachine gun:

- Patent no. 2,403,306, issued July 2, 1946: Firearms Construction
- Patent no. 2,418,906, issued April 15, 1947: Breech Bolt Retracting Means
- Patent no. 2,424,194, issued July 15, 1947: Extensible Shoulder Stock for Firearms

At the time of the applications, the program to further simplify the M3 had already begun. So, only a few months later, some of the patented features, such as the crank-style cocking handle, were eliminated in the series production. Additional patents for the specific characteristics of the simplified model M3A1 were applied for.

A Thompson M1928A1 and a pre-series M3 with a layer of dried mud. They were immersed with inserted magazines. The muzzles were sealed with tape. *APG*

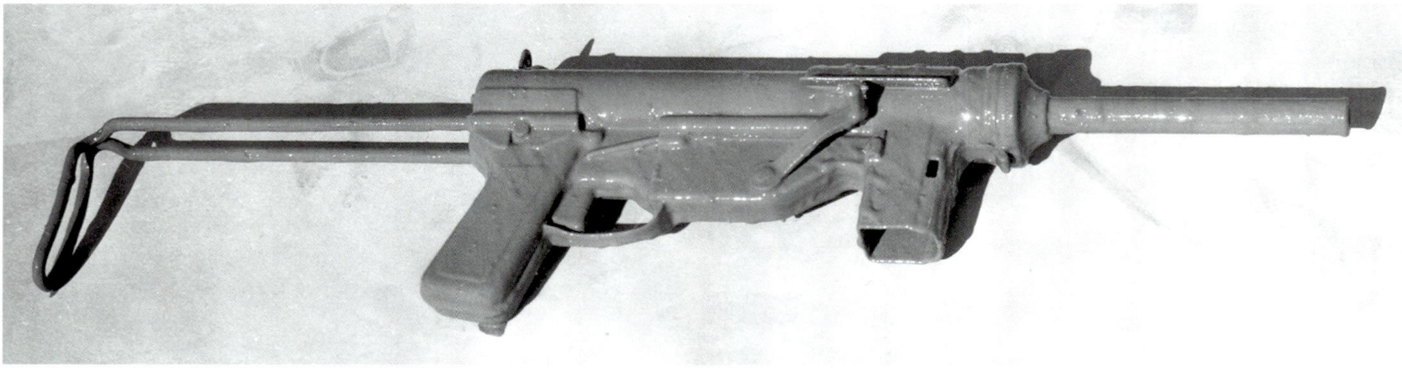

An M3 fresh out of the mud bath. The muzzle was not sealed and no magazine was inserted. *APG*

"New Army Sub-Machine Gun": Cpl. Lloyd W. Ricketts from Fort Meyer, Virginia, is posing for the press. The text on the back says: "One of the new weapons proving so effective against the enemy is the M-3 sub-machine gun, small enough to be carried in an ordinary briefcase, yet so sturdy that after 49,600 rounds of test firing with a single weapon its accuracy improved rather than decreased."

Patented July 2, 1946 **2,403,306**

UNITED STATES PATENT OFFICE

2,403,306

FIREARM CONSTRUCTION

Frederick W. Sampson, Dayton, Ohio, and George J. Hyde, Brooklyn, N. Y., assignors to General Motors Corporation, a corporation of Delaware

Application May 1, 1944, Serial No. 533,569

5 Claims. (Cl. 42—3)

This invention relates to an improved bolt guiding construction for a firearm of the type commonly referred to as a submachine gun or machine pistol.

It is the principal object of this invention to produce a firearm in which the bolt is guided in its reciprocation with accuracy and certainty and one which is not liable to stoppages as a result of exposure to dust, mud, or other foreign elements.

It is a further object of this invention to produce a bolt mechanism assembly for a firearm which may be removed as a unit from the firearm for inspection and cleaning without danger of loss or displacement of springs or other parts.

An additional object is the provision of a bolt guiding construction which requires in its production a minimum of machine work without sacrificing in accuracy and reliability of functioning.

The exact nature of our invention as well as other objects and advantages thereof will be clearly apparent from a study of the following specification referring to the drawings annexed hereto in which:

Fig. 1 is a longitudinal vertical sectional view through a firearm embodying this invention;

Figs. 2, 3 and 4 are cross sectional views on the lines 2—2, 3—3 and 4—4, respectively, of Fig. 1;

Fig. 5 is a longitudinal, sectional view taken substantially along the plane of the line 5—5 of Fig. 1;

Fig. 6 is a pictorial view of the bolt mechanism assembly removed from the receiver.

Referring to the drawings by characters of reference it will be seen that the firearm shown therein is of the same general type as that shown in our copending applications for U. S. Letters Patent, bearing Serial Nos. 533,566; 533,567; 533,568; and 533,570, filed May 1, 1944, and in which the subject matter disclosed but not claimed in this application is claimed, and comprises a receiver 1 to which a barrel sleeve 4 has been welded or otherwise secured. A barrel 5 is threadedly engaged with the barrel sleeve 4 through the medium of the externally threaded barrel bushing 6.

The receiver serves as a housing for the bolt assembly, shown separately in Fig. 6, which comprises a breech bolt 10 formed with an integral firing pin 11 and provided with a spring extractor finger 12 received in a longitudinal bore in the bolt and retained therein by a transverse pin 13. A pair of spaced holes 14 penetrate the bolt longitudinally and are each in substantially parallel relation to the axis of the bolt. A pair of spaced parallel bolt guide rods 15 are fixed in the guide rod retaining plate 16. Surrounding the guide rods and confined thereon between the retaining plate 16 and rearwardly facing surfaces 14*a* on the bolt are a pair of driving springs 17. The guide rods extend forwardly through the bolt guide holes 14 and pass through holes in the guide rod locating plate 18 which engages forwardly facing shoulders 15*a* on each of the guide rods. Grooves 19 in the ends of each of the guide rods forward of the locating plate are engaged by a guide rod retaining clip 20, shown clearly in Figs. 3 and 6.

When the bolt mechanism assembly is inserted into the front end of the receiver lateral projections or locating lugs 21 on the locating plate 18 are engaged in grooves 22 in the barrel sleeve 4 and the rear ends of the guide rods 15 are received in the holes 23 formed in the rear wall of the receiver. When the barrel bushing 6 is screwed into the receiver in front of the bolt assembly the rear face of the bushing engages the front face of the guide rod locating plate and forces the plate, guide rods, and retaining plate back into tight engagement with the rear wall of the receiver. The rear end of the barrel 5 extends through the central aperture 18*a* in the locating plate in position to receive in the chamber 5*a* cartridges fed from the magazine 33. It will be noted that when the parts are secured in this manner the ends of the guide rods and the retaining clip are engaged in an annular space defined by the barrel bushing, the barrel and the locating plate, and that it is impossible for the clip to become loosened or deranged.

It will be readily apparent that by this means the entire bolt and bolt guiding mechanism has been integrated into a compact and convenient subassembly which may be removed from the receiver as a unit and which normally need not be further disassembled for cleaning or inspection. When occasion does arise calling for further disassembly or replacement of parts the retaining clip may be readily removed and the locating plate and bolt allowed to slip off of the guide rods.

Although no detailed description of the lower housing has been included in this application it is believed that it will be evident that the lower housing 44 carrying the ejector 52 and cocking mechanism 55 must be removed before removal of the bolt mechanism can be effectuated. Removal of the lower housing is simply achieved by deforming the resilient trigger guard 49 out of engagement with the hole in the grip and withdrawing the arched end of the guard from the matching holes **47** and **48**, respectively, in the housing and receiver. These features are fully described in our copending application above referred to.

It will be noted that the guide rods furnish complete support and control of the movement of the bolt independently of the walls of the receiver during all phases of bolt travel. This permits the receiver to be formed by stamping or other production methods and without necessitating close tolerances and difficult machining operations on the inner wall of the receiver. At the same time the bolt is guided with a high degree of accuracy, smoother operation is attained, and uniformity of drive spring loading is readily achieved.

This sort of construction presents a minimum of operating and maintenance problems under difficult service conditions since the contacting area between the bolt and its guiding means is greatly reduced as compared to conventional constructions and the bearing area is situated in a well protected position. The spacing which is maintained between the bolt and the wall of the receiver insures that if foreign material should get into the receiver it may not readily jam the mechanism by wedging in a restricted space.

We claim:

1. A firearm comprising a receiver having a substantially closed rear end, locating holes in the rear end, bolt guide rods received in the locating holes, a bolt-slidably mounted on the rods and within the receiver, grooves in the inner surface of the forward end of the receiver, a locating plate having means engaging the grooves and the guide rods, and a barrel bushing engaging the receiver and confining the locating plate and the guide rods.

2. A firearm comprising a receiver, bolt guide rods in the receiver, a bolt slidable on the rods, a centrally apertured guide rod locating plate on the forward end of the rods and engaging the receiver, spring clip means on the rods to secure the plate thereto, a barrel passing through the

central aperture, and a barrel bushing on the barrel engaged with the receiver and with the forward face of the locating plate.

3. A firearm comprising a receiver, a bolt guide assembly in the receiver having a pair of elongated parallel guide rods, a centrally apertured locating plate engaging the receiver and having holes through which the guide rods may pass, opposed grooves in the guide rods, a spring clip exerting resilient outward pressure on the guide rods and engaged in the grooves whereby the plate is secured on the rods, and a barrel assemgly having a portion passing through the central aperture in the locating plate whereby the spring clip is prevented from dislodgment.

4. For use in a firearm a bolt mechanism comprising, a plurality of substantially parallel guide rods having annular grooves adjacent to their forward ends, a retaining plate attached to the rear ends of said guide rods, a bolt including means forming a plurality of longitudinally extending holes therein in which said guide rods are slidably received, driving springs surrounding said rods and operatively engaged for compression between said plate and said bolt, a locating member including a plurality of outwardly extending locating lugs thereon and a plurality of apertures therein receiving the forward ends of said rods, and a resilient retaining clip engaging said grooves forward of said member.

5. For use in a firearm a bolt mechanism comprising, a plurality of guide rods having locking means adjacent their front ends, a retaining plate attached to the rear ends of said rods, a bolt including means forming a plurality of longitudinally extending holes therein in which said rods are slidably received, driving springs surrounding said rods between said plate and said bolt, a locating member in engagement with the forward ends of said rods, and resilient means engaged with said locking means for releasably retaining said rods and said member in engagement.

FREDERICK W. SAMPSON.
GEORGE J. HYDE.

July 2, 1946. F. W. SAMPSON ET AL 2,403,306

FIREARM CONSTRUCTION

Filed May 1, 1944 2 Sheets-Sheet 1

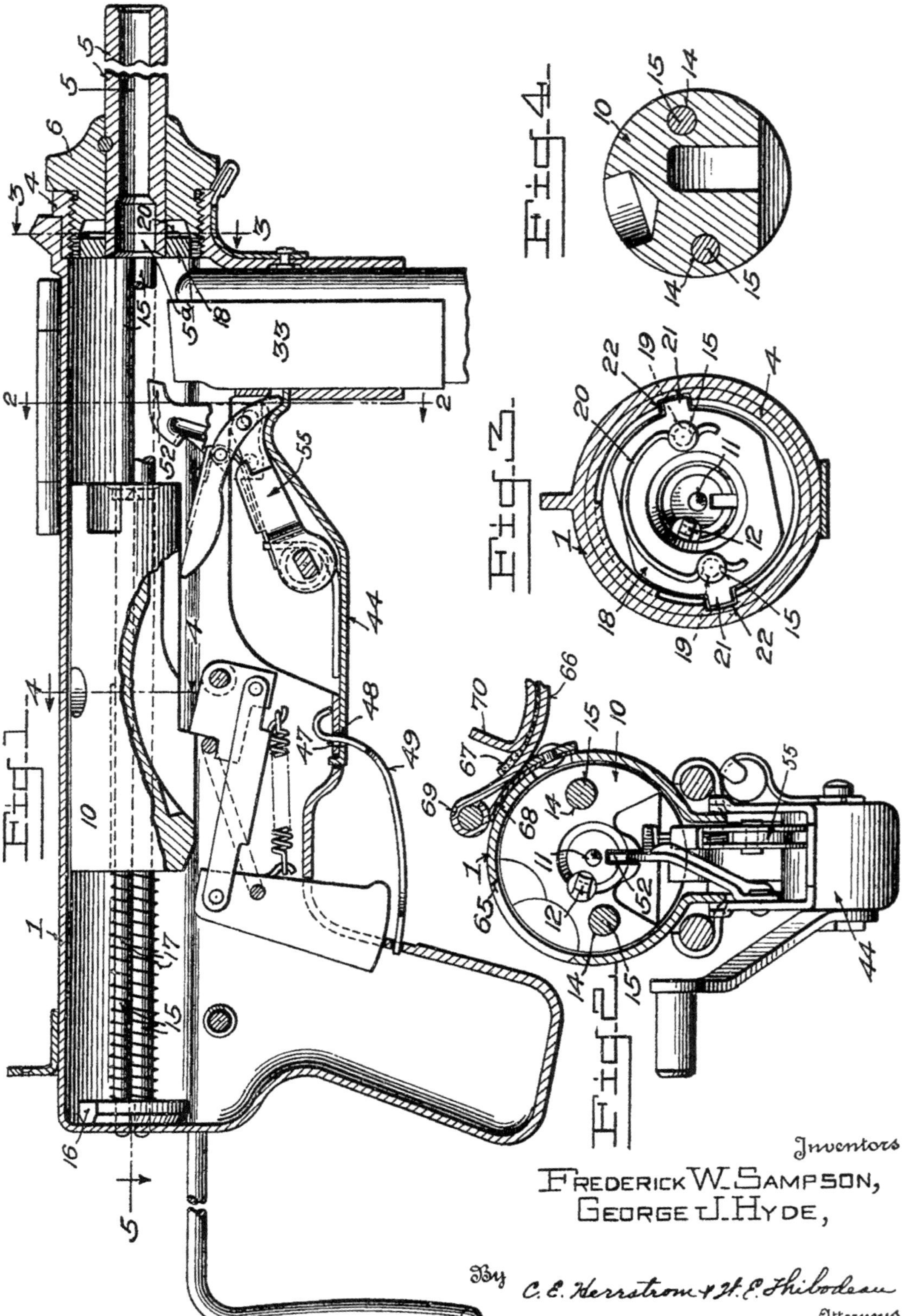

Inventors

FREDERICK W. SAMPSON,
GEORGE J. HYDE,

By C. E. Herrstrom & H. E. Thibodeau
Attorneys

July 2, 1946. F. W. SAMPSON ET AL 2,403,306.

FIREARM CONSTRUCTION

Filed May 1, 1944 2 Sheets-Sheet 2

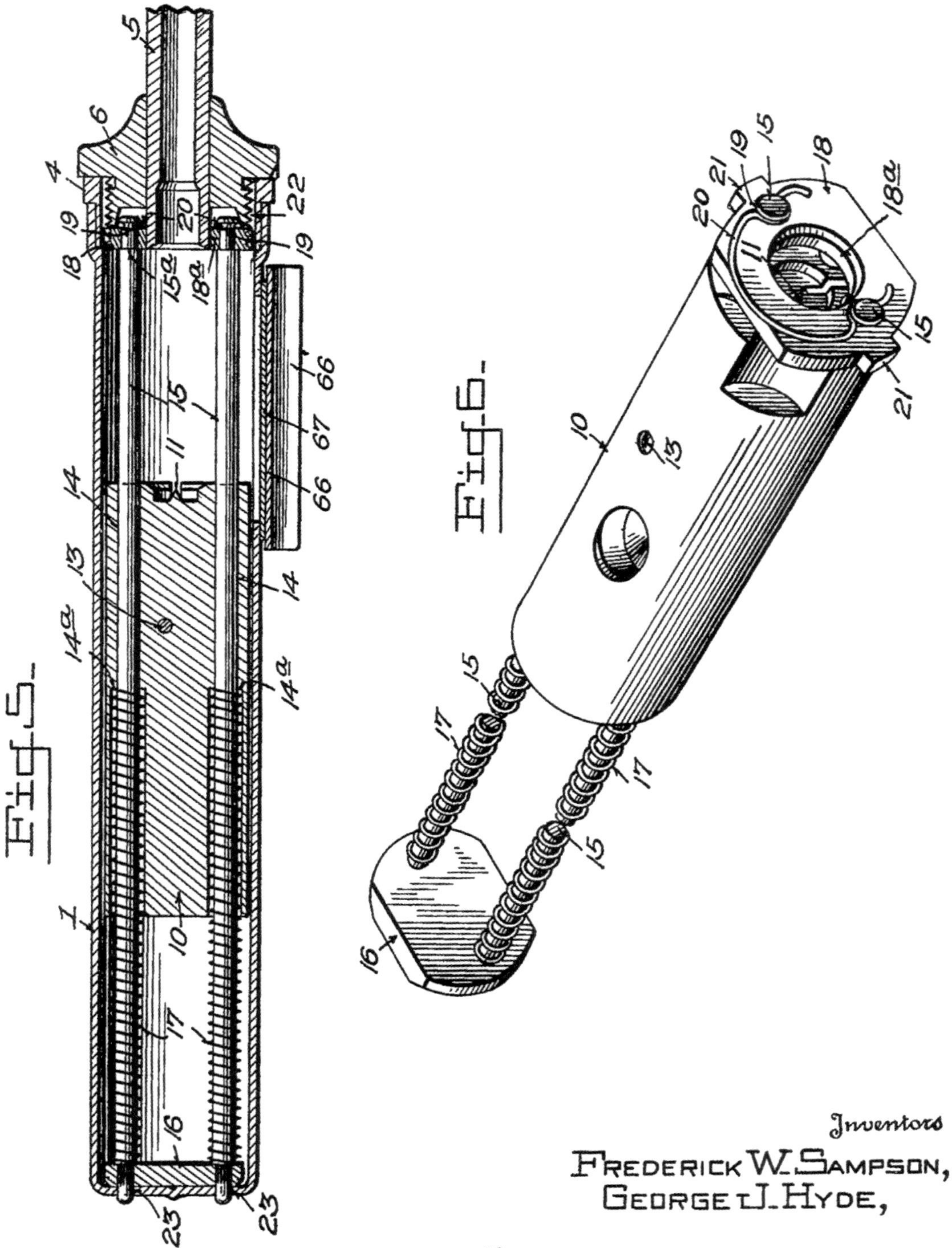

Inventors

FREDERICK W. SAMPSON,
GEORGE J. HYDE,

By C. E. Herrstrom + H. E. Thibodeau

Attorneys

Patented Apr. 15, 1947 2,418,906

UNITED STATES PATENT OFFICE

2,418,906

BREECH BOLT RETRACTING MEANS

Frederick W. Sampson, Dayton, Ohio, and George J. Hyde, Brooklyn, N. Y., assignors to General Motors Corporation, a corporation of Delaware

Application May 1, 1944, Serial No. 533,568

4 Claims. (Cl. 42—16)

1

This invention relates to an improved breech retracting means for a firearm having a reciprocating breech member and more particularly to such means suitable for application to a firearm of the type commonly referred to as a submachine gun or machine pistol.

It is the primary object of this invention to produce a dependable breech retracting means which does not partake of movement of the breech member during automatic fire and does not therefore have any exposed moving parts which might injure hands or clothing or detract from the efficiency of the weapon's operation in confined spaces or in heavy undergrowth.

It is a further object to provide a breech retracting system for firearms which may be mounted and which can be operated without necessitating the provision of openings or slots in the wall of the firearm which can be obstructed by dirt or foreign material.

The exact nature of the invention as well as other objects and advantages thereof will be clearly apparent from an examination of the following specification referring to the annexed drawing in which:

Fig. 1 is a partial, longitudinal sectional view through a firearm embodying our invention;

Figs. 2 and 3 are partial longitudinal sectional views taken in the same plane as Fig. 1 but showing the mechanism in different stages of operation;

Fig. 4 is an exploded pictorial view of the parts making up the cocking mechanism;

Fig. 5 is a cross sectional view on the line 5—5 of Fig. 1.

Fig. 6 is a cross-sectional view on line 6—6 of Fig. 1.

Referring to the drawing by characters of reference it will be seen that the firearm shown therein is of the same general type as that disclosed in our copending applications for U. S. Letters Patent bearing Serial Nos. 533,566; 533,567; 533,569 and 533,570, filed May 1, 1944, in which the structure disclosed but not claimed in this application is claimed, and comprises a receiver 1 having mounted therein a reciprocating bolt 10. The construction shown herein mounts the bolt 10 on spaced guide rods 15 surrounded by driving springs 17 which impel the bolt forwardly into battery position. The particular retracting means to be described hereinafter is not however limited to this construction as it is applicable to the problem of retracting any reciprocating member.

A drawn sheet metal housing 44 is shaped to con-

2

form to the exterior contour of the receiver and is provided at its front end with a tongue 45 engaging a slot in the magazine housing 3 and at its rear end with a tongue 46 engaging a slot in the lower portion of the receiver. Matching apertures 47 and 48 are provided in the housing 44 and in that portion of the receiver opposed thereto. The trigger guard 49 is formed of resilient material and is provided with shoulders 49*a* which will not pass through the matching apertures 47 and 48 and with an arched tongue 50 which does pass through the apertures into contact with the inner wall of the receiver. It will be apparent that when the trigger guard is bent upwardly against its own resiliency into a position in which the tongue 51 may snap into an aperture 2*a* in the grip portion of the receiver that the housing will be securely retained in place. It will be noted that the side walls of the housing overlap the side walls of the receiver and that the lower housing forms a complete closure for the openings in the lower portion of the receiver. In addition to serving as a closure for the lower part of the receiver the housing forms a support for the retracting mechanism, to be described in detail below and the ejector 52 which is riveted or otherwise secured thereto and extends upwardly into a position to be received in a groove 53 in the bolt in position to engage the head of a cartridge gripped by the extractor and impel same upwardly and to the right as the bolt recoils upon firing. A stop tongue 54, cooperating with the retracting mechanism in a manner to be described hereinafter, it cut from the rear edge of the ejector and turned inwardly across the housing.

A breech retracting handle 55 is provided with an axle 55*a* pivotally mounted in the lower portion of the housing 44. Within the housing a portion of the axle is flattened and carries thereon a torsion spring 56 and a retracting lever arm 57 which so engages the flattened portion of the axle that it is constrained to rotate therewith, and under the influence of the spring 56 both the handle and the retracting lever arm tend to swing forwardly in the direction of the magazine housing 3 until such motion is restrained by contact of the lever arm with the bottom wall of the housing 44.

A retracting pawl 58 is pivotally supported on the end of the retracting lever arm and a pawl spring 59 is provided tending to swing the pawl upwardly into a direction substantially in prolongation of the lever arm. This tendency is, however, limited by the presence of a stop 60 near the end of the retracting lever arm 57 which is in position to engage the heel 61 of the pawl and prevent damage to the spring in the event of displacement of the retracting handle from its idle position during intervals when the housing may have been removed from the receiver. It will also be noted that the heel of the pawl engages the arched cam surface 44a as the retracting lever arm nears its forward position, thus rocking the pawl downwardly and to the rear about its pivot. Obviously in order to accomplish this result the spring 56 will have to be capable of exerting greater force than the spring 59. This camming action positively removes the pawl from the path of the bolt during automatic operation of the arm. The tongue 54 on the ejector extends across the path of the pawl and aids the cam surface 44a in removing the pawl from the path of the bolt and in preventing the pawl from assuming a position in which improper assembly might be possible when the housing is attached to the receiver.

As the retracting handle is swung rearwardly against the action of the spring 56 the spring 59 will urge the pawl upwardly into the groove 62 in the lower face of the bolt and into a position in which the toe 63 of the pawl engages a depression 64 in the rear end of the groove. Further retraction of the handle will result in carrying the bolt rearwardly with the pawl into a position in which the sear 26 engages and retains the bolt. Releasing the retracting handle will permit the handle, lever arm, and pawl to return to their inactive forward position in which there is no connection with the bolt and in which position the cocking mechanism may remain until such time as it is again necessary to cock the mechanism manually.

It will be noted that the mechanism shown herein does not present any exposed parts which partake of movement of the bolt during automatic fire and that there is therefore no danger that hands or clothing will be injured during automatic fire. At the same time however, there is provided a most convenient cocking handle which by an easy natural movement will engage itself with and cock the mechanism.

It will also be noted that this mechanism functions without requiring the provision of openings, slots, or track guided members which might permit the entrance of dirt into the interior mechanism or which might in themselves be obstructed by dirt or foreign materials.

Although this mechanism finds its greatest utility as a breech retracting means for an automatic firearm it will be obvious that similar means might be applied to the intermittent manual operation of any reciprocating member.

We claim:

1. In a firearm, the combination of a reciprocating firearm member, a lever journaled in the firearm, a pawl member having a pawl portion and an extension portion in longitudinal alignment therewith, means intermediate said portions pivotally mounting said pawl member on said lever, means causing said pawl to engage the firearm member when the lever is actuated in one direction and means engaging said extension portion when said lever is moved in the reverse direction whereby said pawl is drawn clear of the path of the firearm member.

2. A firearm cocking mechanism comprising a housing, a lever journaled in the housing, means to swing the lever, a pawl member having a pawl portion and an extension portion longitudinally in alignment therewith, means intermediate said portions pivotally mounting said pawl member on said lever, a spring abutting said lever tending to bias the same to an idle position, spring means tending to swing the pawl member toward a position in substantial prolongation of the lever, and means on the housing engaging said extension portion as the lever approaches its idle position to cause said pawl portion to assume a pivoted back position on the lever when the lever is in idle position.

3. In a firearm, the combination of a member to be cocked, a housing, a crank, a spindle fixed to the crank and journaled in the housing, a lever arm within the housing and engaging the spindle to be rotated thereby, a spring engaging the housing and the lever arm and tending to bias the crank and the lever arm to a forward idle position, a pawl member having a pawl portion and an extension portion in longitudinal alignment therewith, means intermediate said portions pivotally mounting said pawl member on said lever-arm, spring means engaged between the pawl portion and the lever arm and biasing the pawl member toward a position for engagement with the member to be cocked, and means on the housing engaging said extension portion when the lever is in idle position to hold the pawl portion away from engagement with the member to be cocked.

4. In an automatic firearm, the combination of a reciprocating firing member, a cocking mechanism therefor including a pawl member having a pawl portion and an extension portion in longitudinal alignment therewith, a crank means operatively connected to said pawl member intermediate said pawl portion and said extension portion, means causing said pawl portion to engage said firing member to cock the same when the crank means is moved in a firearm cocking direction, and means arranged to engage said extension portion to remove said pawl portion from the path of said firing member when said crank means is moved in a direction opposite to said cocking direction to a non-operating position.

FREDERICK W. SAMPSON.
GEORGE J. HYDE.

REFERENCES CITED

The following references are of record in the file of this patent:

UNITED STATES PATENTS

Number	Name	Date
785,085	Burton	Mar. 21, 1905
1,003,790	Pordon et al	Sept. 19, 1911
237,357	Zundorff	Feb. 1, 1881
1,057,171	Redfield	Mar. 25, 1913
1,472,126	Kewish	Oct. 30, 1923
534,894	Laumann	Feb. 26, 1895

FOREIGN PATENTS

Number	Country	Date
10,391	British	1900
389,737	German	Feb. 7, 1924

April 15, 1947. F. W. SAMPSON ET AL 2,418,906

BREECH BOLT RETRACTING MEANS

Filed May 1, 1944

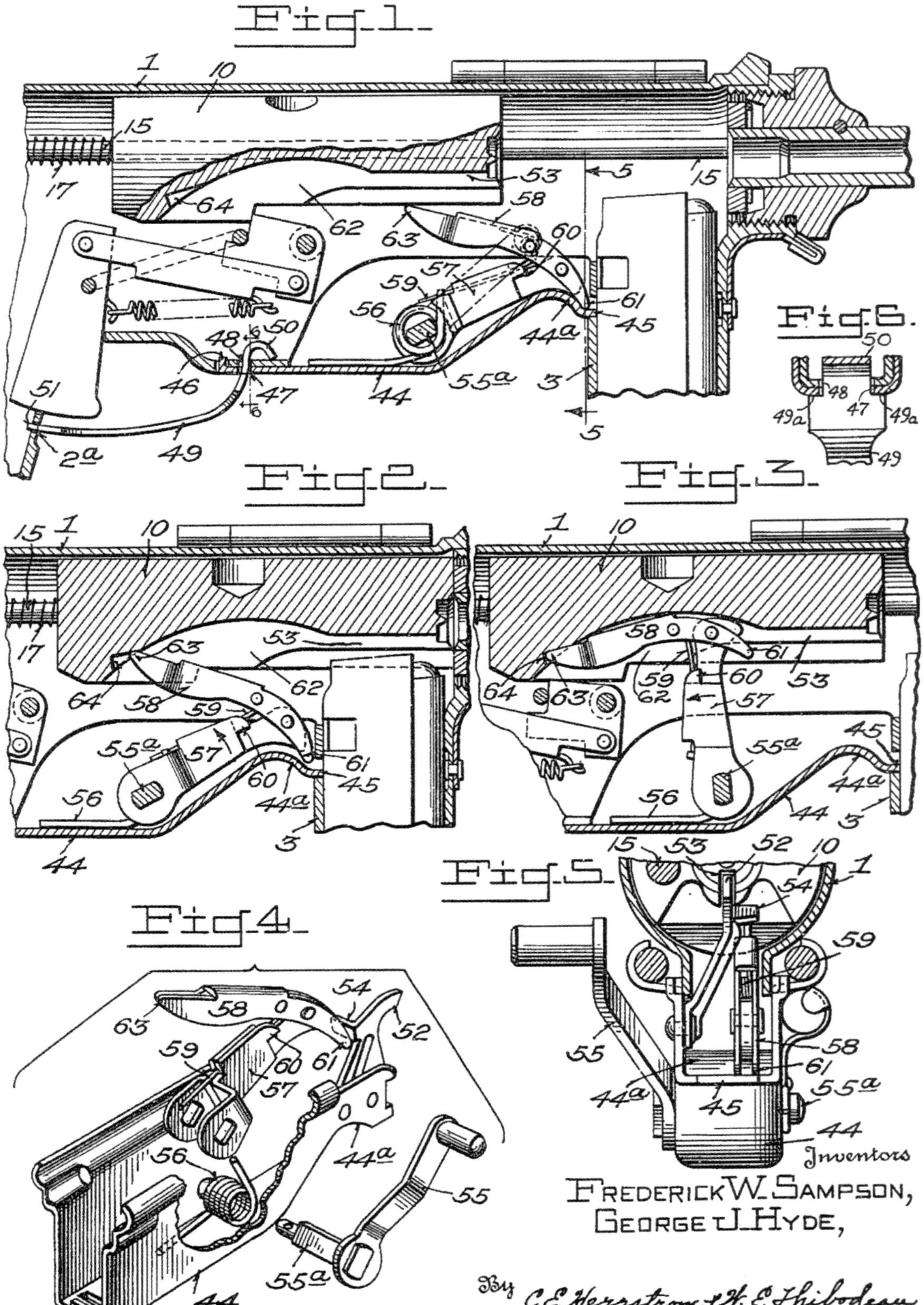

Patented July 15, 1947

2,424,194

UNITED STATES PATENT OFFICE

2,424,194

EXTENSIBLE SHOULDER STOCK FOR FIREARMS

Frederick W. Sampson, Dayton, Ohio, and George J. Hyde, Brooklyn, N. Y., assignors to General Motors Corporation, a corporation of Delaware

Application May 1, 1944, Serial No. 533,570

4 Claims. (Cl. 42—72)

1

This invention relates to a stock for a firearm and more particularly to a collapsible stock for a firearm of the submachine gun or machine pistol type.

The primary object of this invention is the production of a stock which has sufficient strength and rigidity for practical use and which may be collapsed to occupy a minimum of shipping and carrying space.

It is a further object to produce a firearm stock member which has a minimum weight consistent with strength.

An additional object is the production of a firearm stock member which may be readily removed to facilitate compact packaging and which may be, when removed, used as a convenient cleaning rod for the firearm.

Another object is the provision of a firearm stock which may be used and secured in positions of varying extensibility.

All of the objects above mentioned and those which may appear hereinafter may be related directly to the problem of producing a weapon suitable for military use and particularly in reference to the problem of producing a satisfactory arm for use by paratroops and similar organizations.

The exact nature of the invention as well as other objects and advantages thereof will be apparent from an inspection of the following specification referring to the annexed drawing in which:

Fig. 1 is a side elevational view of a firearm embodying our invention;

Fig. 2 is a rear elevational view corresponding to Fig. 1;

Figs. 3 and 4 are partial cross sectional views on the line 3—3, Fig. 4 showing the latch in the position it would occupy with the stock in the dotted line position of Fig. 1;

Fig. 5 is a bottom plan view of the stock removed from the receiver;

Fig. 6 is a partial, side elevational view of the stock, partially in section to show the adaptability of the stock to use as a cleaning rod.

Referring to the drawing it will be seen that the firearm to which our invention has been applied is of the same general type as that disclosed in our copending application bearing Serial Numbers 533,566, 533,567, 533,568, 533,569, filed May 1, 1944, and in which the structure disclosed but not claimed in this application is claimed.

The stock **77** comprises a continuous rod member which has been bent to define a shoulder engaging portion **77***a* and a pair of straight, parallel, forwardly extending rod members **78** which are received in slidable relation in stamped ways **79** welded or otherwise secured to the receiver. Guide ways **80** are also provided on each side of the housing **44** to receive the ends of the stock rods **78** when it is fully telescoped with relation to the receiver. The stock structure may be conveniently stiffened somewhat by the provision of a welded or similar joint in the region **77***b* near the junction of the shoulder engaging portion and the parallel rods.

2

A locking catch **81** is mounted in the receiver for transverse sliding movement and is normally impelled to the left by the spring **82** which engages between the inner wall of the receiver **1** and the cylindrical latch portion **81***a* which is conveniently provided with a finger button **81***b*. A similar cylindrical latch portion **81***c* is conveniently riveted on the catch **81** serving both as a latch and by engagement with the wall of the receiver as a limiting stop for the catch. Notches **83** are provided on the stock rods and are engaged by the cylindrical latches **81***a* and **81***c* to releasably retain the stock in short position. The fact that these notches are formed with a rounding face and do not pass entirely across the rod makes it possible to apply a deliberate pull to the stock and extend same without releasing the latch by a separate movement, as the latches may be wedged out of the notches by the rounding face to permit the extension of the stock.

It will likewise be readily apparent that the latch portions **81***a* and **81***c* will frictionally engage the rods **78**, when the stock **77** is telescoped or retracted, to resist extension thereof and to hold the stock thus positioned for all practical purposes.

Semicylindrical notches **84** are provided near the ends of the stock rods for engagement with the latches when the stock is fully extended. It will be noted that these notches pass entirely across the rods and that when the latches are engaged therewith the stock cannot be moved except upon deliberate and complete depression of the latch button. It will also be noted that ramp surfaces **84***a* are provided leading into the notches **84**. These ramp surfaces insure that if the stock is extended rapidly there will be no opportunity for the latches to miss engagement with the notches and thus possibly permit an inadvertent removal of the stock. In the event that it is desired to removed the stock completely this can readily be accomplished by deliberate and complete depression of the catch button 81b which will release the stock for withdrawal from the guide ways 79.

When the stock has been removed from the receiver it will be noted that the parallel rod portions of the stock are ideally adapted for use as cleaning rods. To facilitate this use the ends of the rods are threaded internally to receive cleaning implements such as the brush 91.

It is believed that it will be obvious that by this means a collapsible stock has been provided which when extended provides rigid and comfortable support for firing the weapon from the shoulder. It will be noted that the line of reaction from the recoil forces is very nearly in line with the points at which those forces must be absorbed. When applied to a submachine gun of the type on which it is shown this results in a negligible tendency to climb during the firing of long bursts since the force couple due to recoil forces is more than balanced by the weight of the arm.

It will also be noted that the stock can be telescoped to take up a negligible amount of space for carrying or stowage and that with the stock in this position or removed entirely the weapon may be used as a pistol or fired from the hip without difficulty.

All of these factors are of considerable importance as applied particularly to military use by air borne combat forces. While the invention has been illustrated as applied to an automatic submachine gun it is believed that it will be recognized that such a stock could be applied to firearms of any size and type and that it might be conveniently applied to arms of the sporting type for use in densely wooded areas or under other conditions where it might be desirable to travel light and pack compactly.

1. A firearm comprising a receiver, guideways on and at opposite sides of the receiver and extending parallel to the central longitudinal axis thereof, an extensible shoulder stock including spaced parallel bars slidably engaging said guideways, respectively, a latch movably mounted on the receiver to reciprocate substantially along the intersection of a plane normal to said receiver axis with a plane adjacent and parallel to the plane of said guideways, said bars having notches in complementary portions thereof, and said latch

having spaced latch portions constructed and arranged to engage said notches to retain the stock in a definite position relatively to the receiver.

2. A firearm as in claim 1, and a spring carried by the latch and bearing against a portion of the receiver and a portion of the latch to yieldably retain said latch portions in engagement with the notches.

3. A firearm as in claim 1, a spring carried by the latch and bearing against a portion of the receiver and a portion of the latch to yieldably retain said latch portions in engagement with the notches, said bars having ramp portions formed therein and opening into the notches, for receiving and directing said latch portions toward the notches, as the stock is moved toward and approaches said aforementioned position.

4. A firearm comprising a receiver, guideways located at the sides of the receiver, an extensible shoulder stock including spaced parallel bars slidably engaging said guideways, a latch movably mounted in and transversely of said receiver, said bars having notches in complementary portions thereof, said latch having spaced latch portions for engaging said notches to retain the stock in a definite position relatively to the receiver, and a spring carried by the latch and bearing against a portion of the receiver and a portion of the latch to yieldably retain said latch portions in engagement with the notches, said bars having a second set of notches constructed and arranged to receive the latch portions when the stock is in another position relatively to the receiver, and said latter notches having rounded ends for camming the latch portions out of engagement therewith.

FREDERICK W. SAMPSON.
GEORGE J. HYDE.

REFERENCES CITED

The following references are of record in the file of this patent:

UNITED STATES PATENTS

Number	Name	Date
2,200,780	Schmeisser	May 19, 1940
202,946	Johnson	Apr. 30, 1887
593,890	Houston	Nov. 16, 1897
562,487	Quackenbush	June 23, 1896
1,877,016	Munson	Sept. 13, 1932

July 15, 1947. F. W. SAMPSON ET AL 2,424,194

EXTENSIBLE SHOULDER STOCK FOR FIREARMS

Filed May 1, 1944

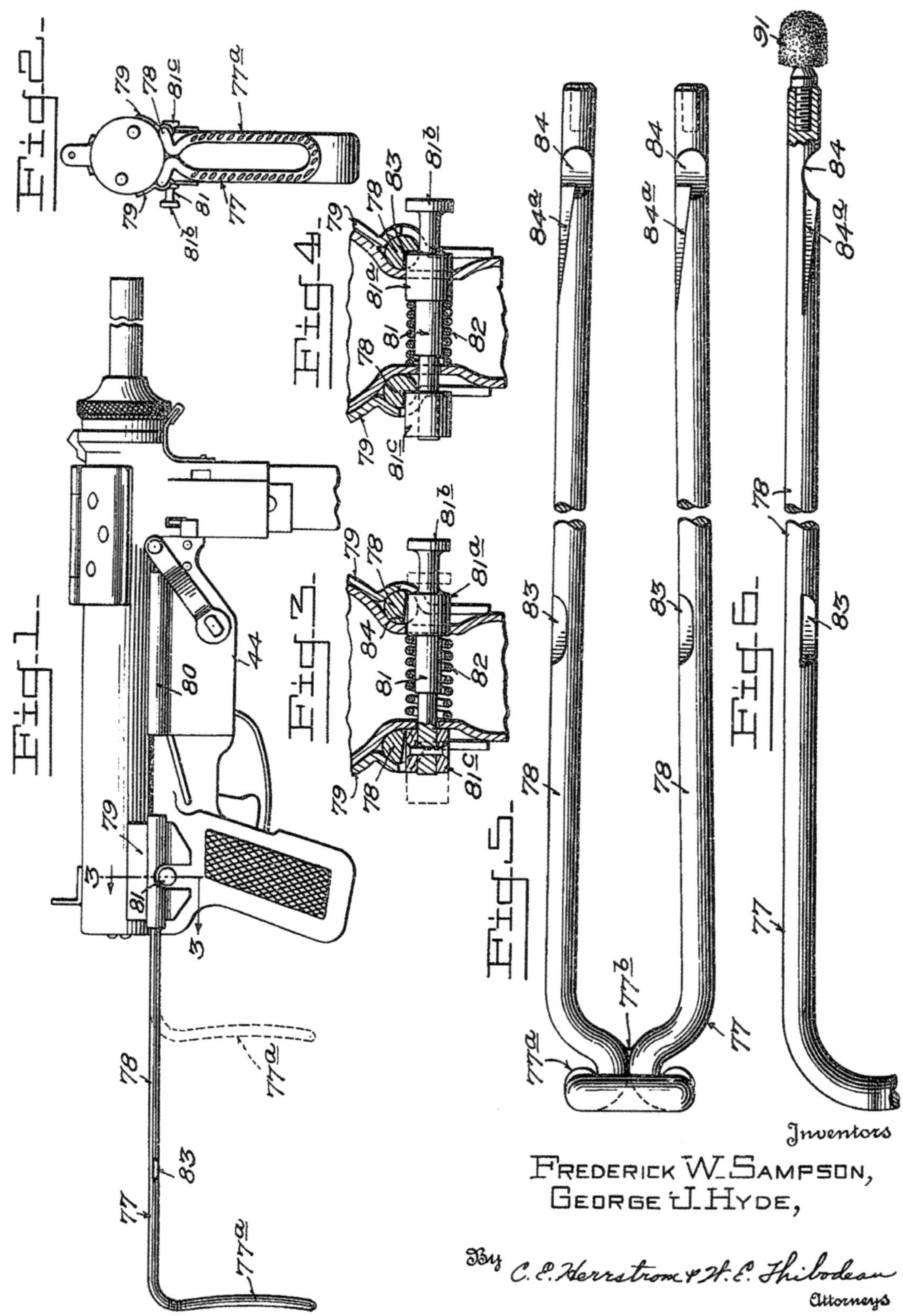

Inventors

FREDERICK W. SAMPSON,
GEORGE J. HYDE,

By C. E. Herrstrom & W. E. Thibodeau
Attorneys

05 THE M3 IN DETAIL

THE "U.S. SUBMACHINE GUN, Caliber .45, M3" is an automatic, air-cooled, blowback-operated weapon that fires from an open bolt. The weapon works in full automatic only, but thanks to the low rate of fire of only 400 to 450 rounds per minute, a skilled operator can discharge single shots.

The bolt is cocked by using a crank-type cocking handle, located on the right side of the ejector housing. After insertion of a full magazine, the spring-loaded crank is rotated to the rear and then released. During this process, a pawl rises to engage a notch in the bottom of the bolt, pushing the bolt to the rear until it locks back on the sear. When the trigger is pulled, the bolt is driven forward by the recoil springs, stripping a round from the feed lips of the magazine and guiding the round into the chamber. The bolt then continues forward and the firing pin strikes the cartridge primer, igniting the round, resulting in a high-pressure impulse, forcing the bolt back against the resistance of the recoil springs and the inertial mass of the bolt. By the time the bolt and empty casing have moved far enough to the rear to open the chamber, the bullet has left the barrel, and pressure in the barrel has dropped to a safe level. The crank is not permanently connected to the bolt and does not move during firing.

The bolt is drilled longitudinally to support two parallel guide rods, upon which are mounted twin driving (recoil) springs. This configuration allows more-generous machining tolerances while providing operating clearance in the event of dust, sand, or mud ingress. A spring-loaded extractor is housed inside the bolt head. The two parallel guide rods are connected to one another with plates at the front (guide rod locating plate) and rear end (guide rod retaining plate), so that they are kept in their position inside the weapon's receiver.

The weapon's only safety is the hinged ejection port dustcover; a projection is riveted to the underside of the cover. When closed, it engages a notch on the bolt, locking it either in its forward or rearmost position. Before the weapon is fired, the dustcover has to be opened by hand. The M3 has no mechanical means of disabling the trigger.

The design of the rear sight was changed shortly after the start of series production. The early type was made from an L-shaped metal strip with an aperture. However, this was prone to damage and had to be strengthened by adding lateral gussets. It is zeroed for 100 yards and not adjustable. The blade foresight, located in front of the dustcover, is formed with each receiver half of one piece of sheet metal and then welded together. All M3 submachine guns were test-fired at the factory for accuracy at a distance of 100 feet.

The weapon can be fired both with retracted and extended stock. The stock is one piece of formed steel rod, which can be telescoped into tubes on both sides of the receiver. It is not possible to adjust the stock in steps; it can be locked only in the forward or rear position. Both ends of the stock are tapped and drilled so that they can be used as cleaning rods. The stock can also be used as a wrench to unscrew the barrel nut or as a disassembly tool to remove the spring steel trigger guard.

Test of the M3 with reinforced rear sight

A "U.S. Submachine Gun M3" of early series production. The L-type rear sight, made from sheet metal, was prone to deformation.

The "hook" riveted to the underside of the cover engages a notch on the bolt. That is the only safety mechanism. The "H" on the bolt means "hardened."

The new rear sight with lateral gussets was much more stable.

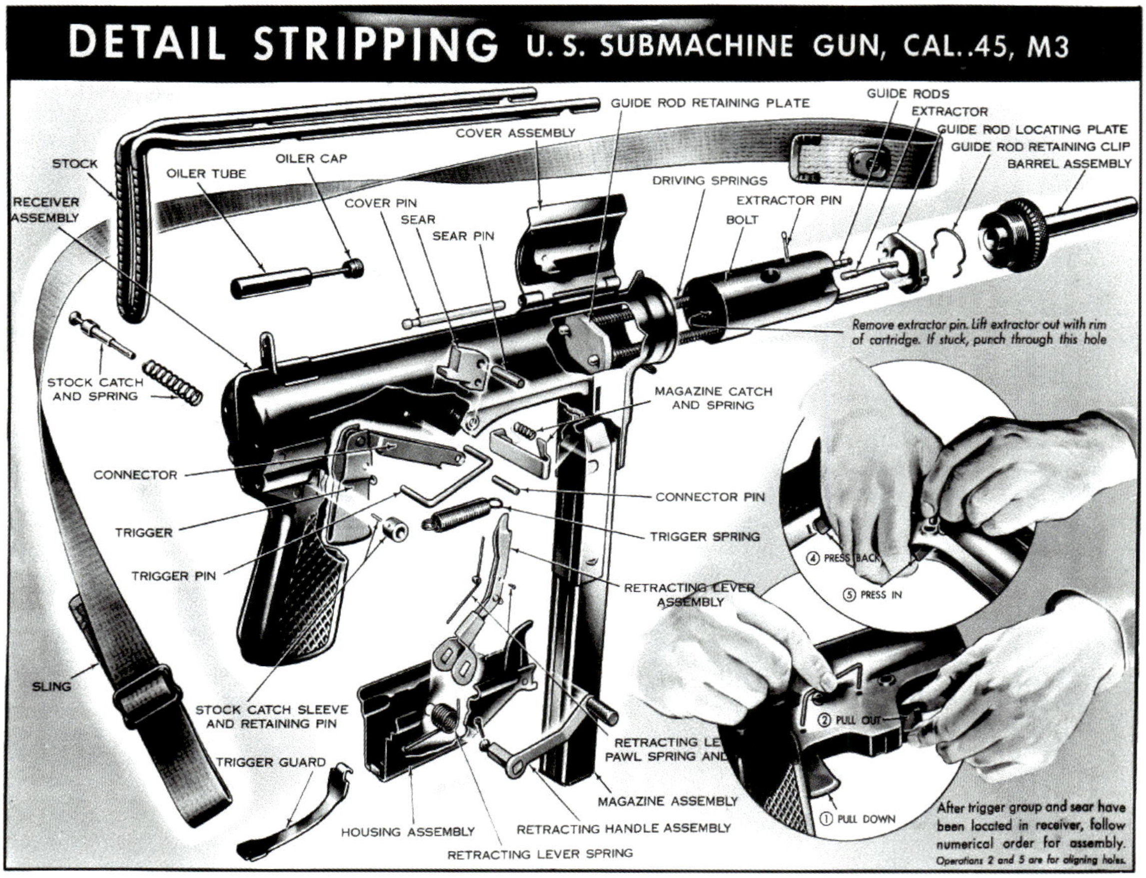

This illustration from US Army training materials is showing the designations of individual parts.

M3 TECHNICAL DETAILS

Caliber	.45 ACP
Weight (empty)	8.15 lbs. (3.7 kg.)
Length	22.8 in. (stock retracted)
	29.8 in. (stock extended)
Barrel length	8 in. (4 grooves / right-hand twist)
Rate of fire	400–450 rounds per minute
Magazine capacity	30 rounds
Sight	rear peep sight for 100 yds.

DISASSEMBLING THE M3

The submachine gun M3 can be disassembled completely without using tools. This makes troubleshooting and the replacement of defective parts very easy.

First, press in the stock catch on the left side of the pistol grip and remove the stock by pulling it directly to the rear.

To remove the trigger guard, place one side of the shoulder rest of the stock on the housing assembly, against the trigger guard,

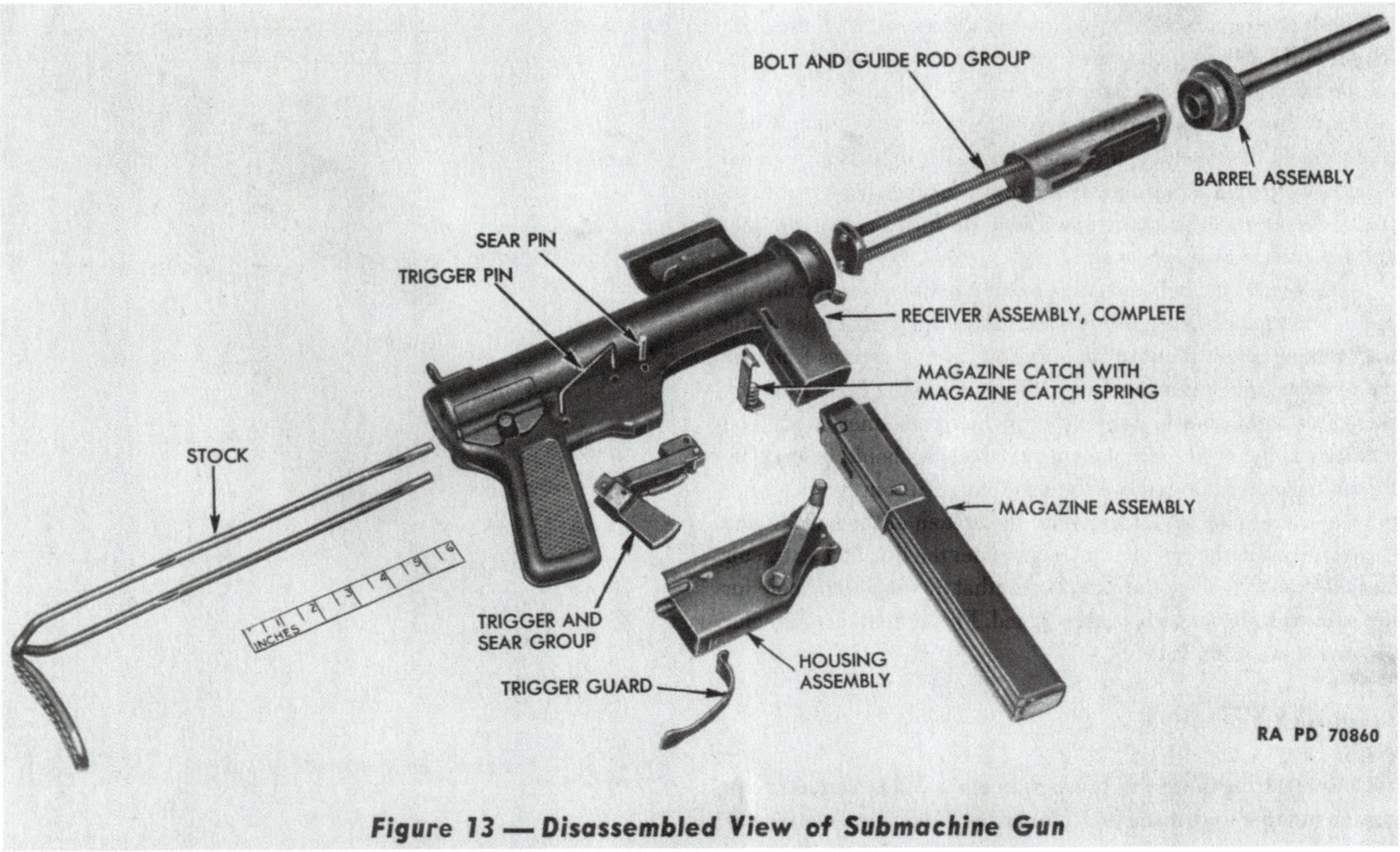

Disassembled M3 according to Technical Manual 9-217, issued in July 1943

The M3 can be disassembled without any tools. Only tongs are needed to remove the guide rod retaining clip when taking apart the bolt assembly.

and pry the spring steel trigger guard out of the pistol grip. Rotate the trigger guard toward the front of the weapon, and unhook the trigger guard from the housing assembly. Remove the housing assembly by pulling up and to the rear on the rear end.

Remove the magazine catch assembly by rotating it toward the right side of the receiver.

To remove the barrel, make sure the bolt is forward, depress the barrel ratchet, and unscrew the barrel. To avoid unnecessary wear, do not allow the barrel ratchet to contact the notches in the barrel collar while screwing the barrel off and on.

Open the dustcover and withdraw the bolt and guide rod assembly from the receiver. Remove the trigger pin. Withdraw the trigger and sear group from the receiver (be careful not to drop the connector pin). This step completes field disassembly.

Any further disassembly wasn't necessary for cleaning or replacement of defective parts, and should be, if possible, carried out only by trained personnel; the cocking mechanism was especially tricky. As a result of careless handling, the springs could be bent and thus cause malfunctions.

Disassembly of bolt and guide rod group: Compress the driving springs and remove the guide rod retaining clip. Remove the guide rod locating plate. Remove the bolt and driving springs from the guide rods. Do not remove the guide rod retaining plate from the two guide rods (this is done by ordnance personnel only). The extractor is removed for replacement only (this should be done by the unit armorer). To remove the extractor, drift out the extractor pin (from the small end, located on the bottom of the bolt). Place a screwdriver or the rim of a cartridge under the lip of the extractor and lift it out. Replace the extractor so that the notch and hole for the extractor pin in the bolt are aligned. Firmly seat the extractor pin and stake it in place

THE M3 IN 9 MM CALIBER

Sometime during the series production of the M3 at Guide Lamp, a small number were made in 9 mm caliber. No documentation has yet been found, and the circumstances are still a mystery. However, some of these weapons are still in existence. At first glance, they look like a normal M3 with the reinforced rear sight. But the magazine well is marked only with "U.S. 9MM S.M.G."

SUBMACHINE GUN, CAL. .45, M3

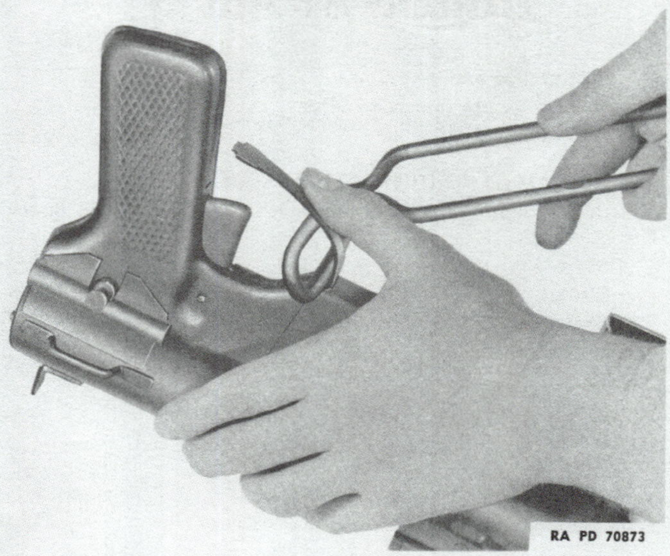

Figure 9 — Removal of Trigger Guard

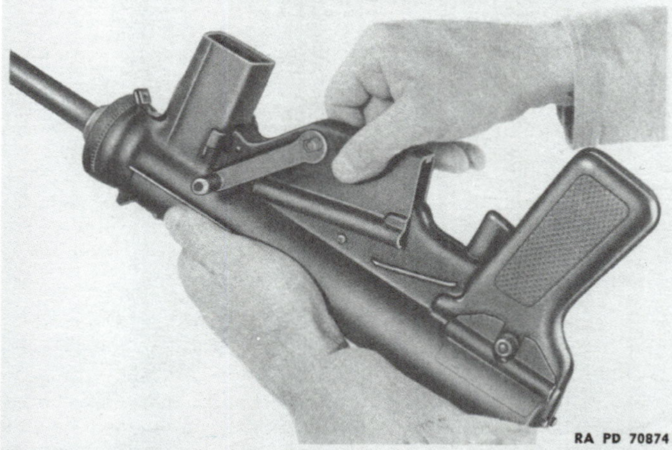

Figure 10 — Removal (or Replacement) of Housing Assembly

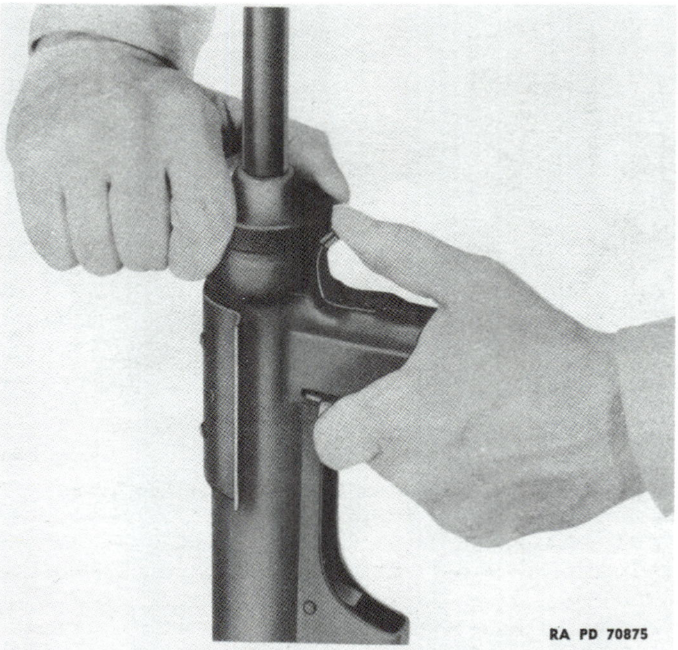

Figure 11 — Removal of Barrel Assembly

Top right: The wire stock serves as a tool for easier removal of the trigger guard (figures 9 and 10 from TM 9-217).

Top left: The barrel nut is held in place by a bent strip of spring steel (figure 11 from TM 9-217).

After removal of the barrel assembly the bolt can be pulled out of the receiver (figure 12 from TM 9-217).

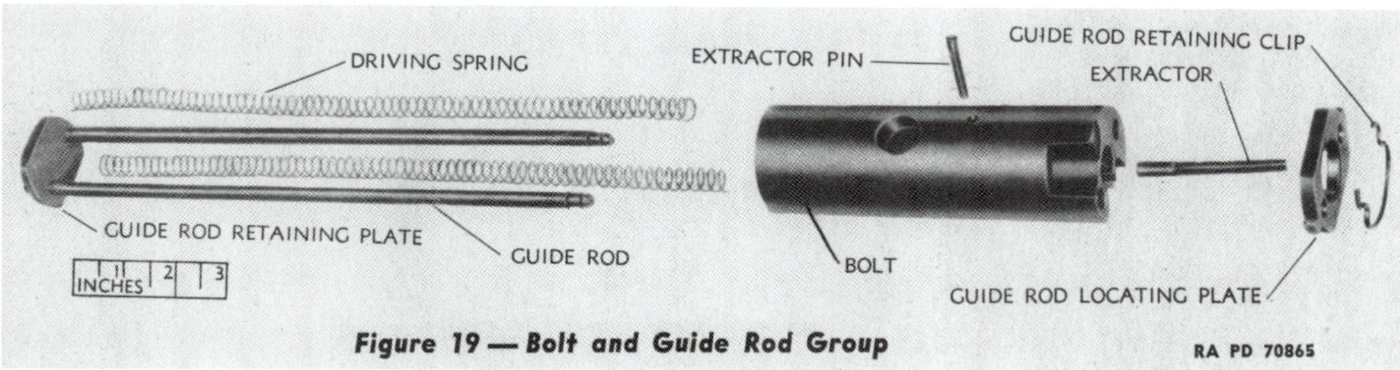

A disassembly of the bolt and guide rod group was not necessary except for purposes of repair (figure 19 from TM 9-217).

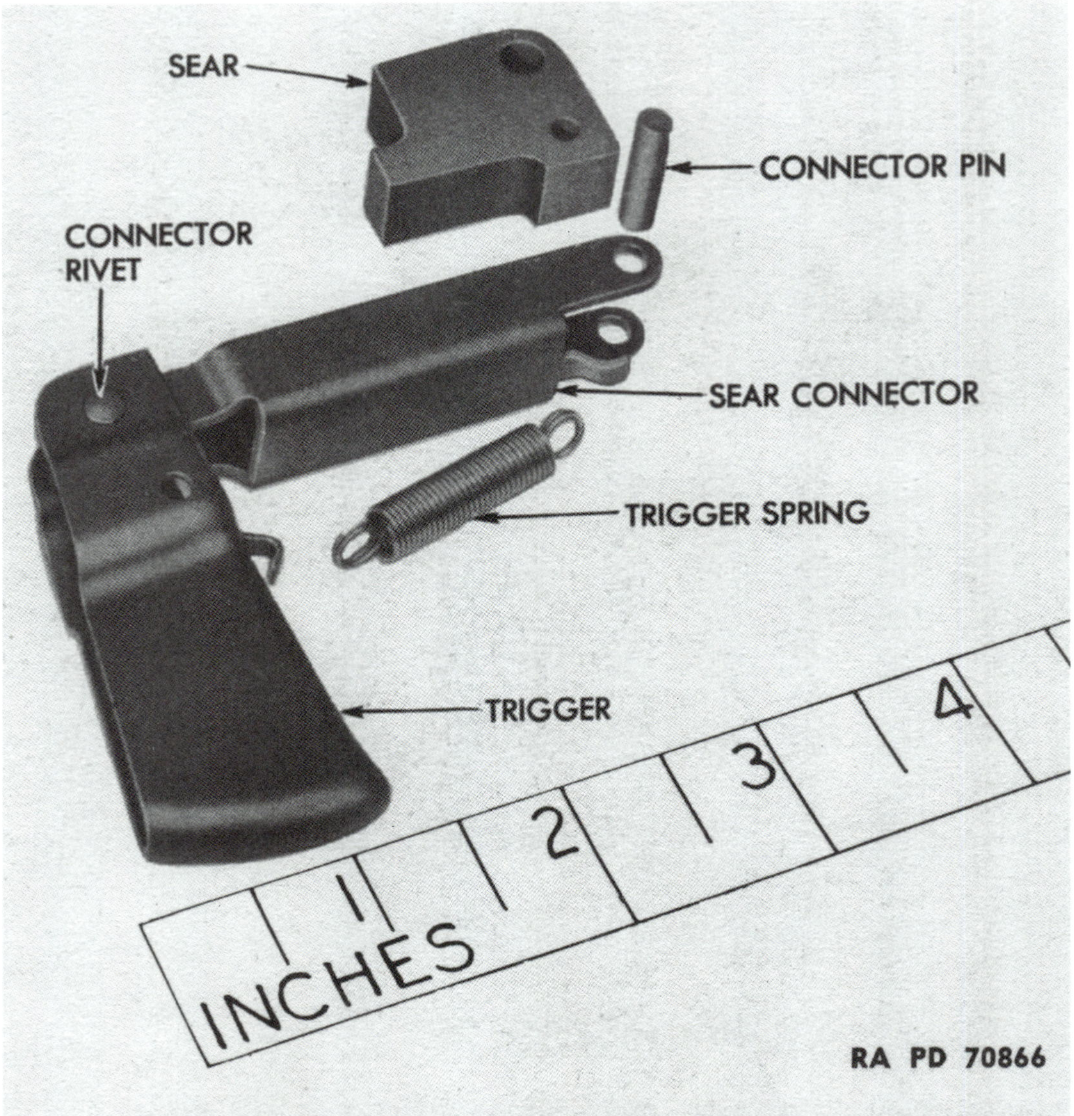

Figure 20 — Trigger and Sear Group

The trigger assembly consists of very few parts (figure 20 from TM 9-217).

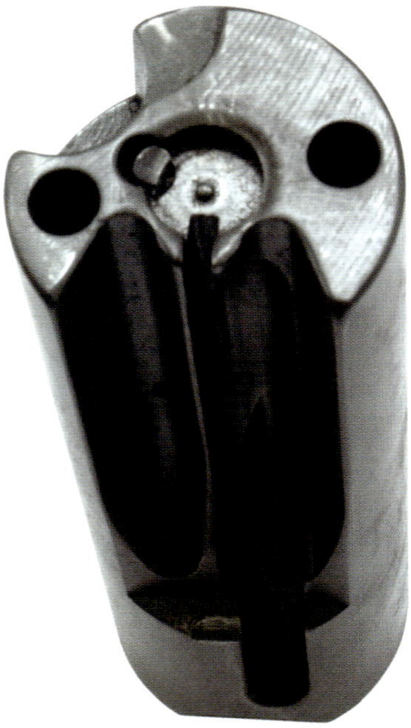

Left: The firing pin was milled into the bolt face and is not replaceable. *Courtesy of Frank Iannamico*

Right: Trigger assembly removed from the weapon, showing the reinforced sear as introduced in the early 1950s

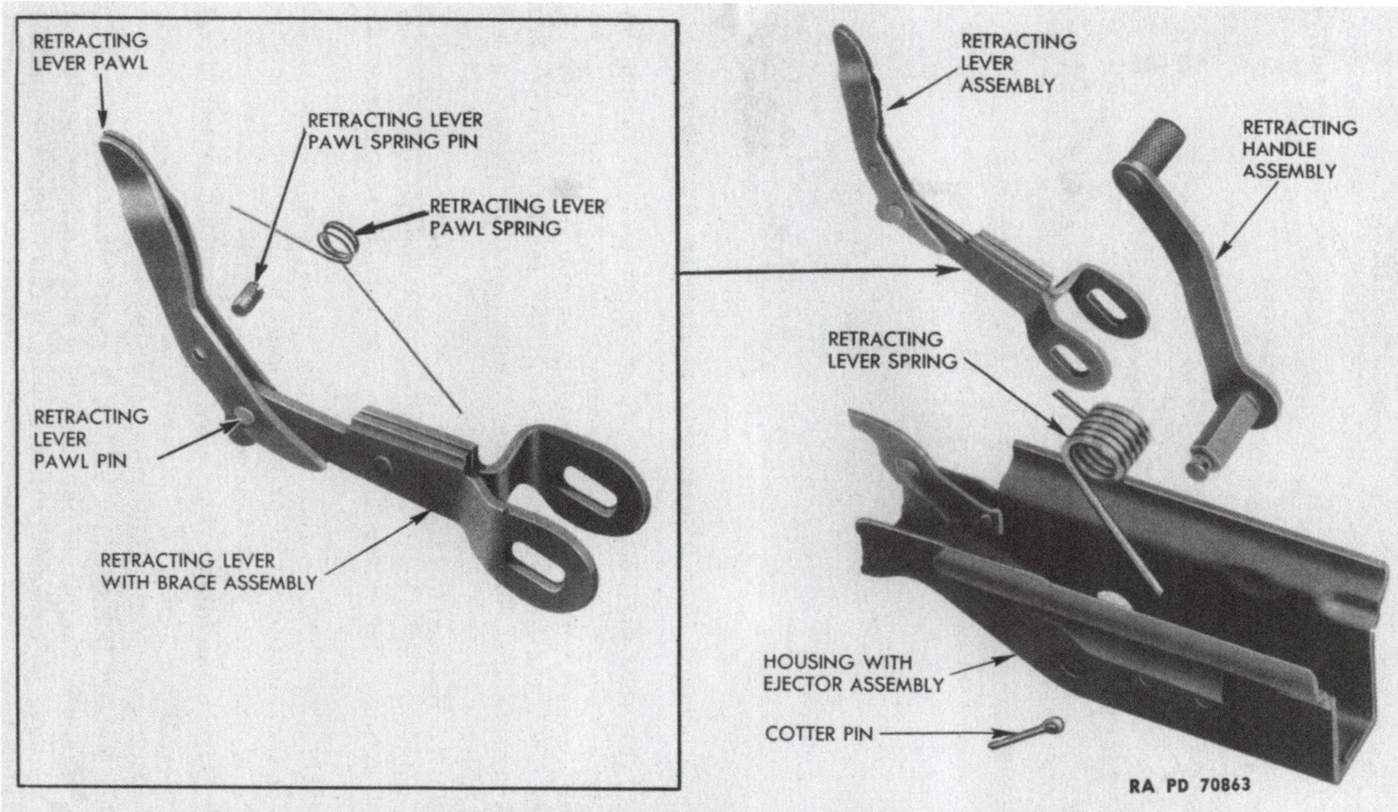

Figure 16 — Housing Assembly

One of the rare M3s that left the factory with the standard 9 mm kit already installed. *Courtesy of the Swiss Military Museum, Full-Reuenthal*

The only marking on these weapons is "U.S. 9MM S.M.G." *Courtesy of the Swiss Military Museum, Full-Reuenthal*

Detail of the magazine well adapter. *Courtesy of the Swiss Military Museum, Full-Reuenthal*

06 FROM THE M3 TO THE M3A1

AS SO OFTEN HAPPENS, minor problems with a new weapon will show up during its use. Despite the intensive pretesting, the first technical difficulties of training units in the United States had been reported back in February 1944. During hard use in the ongoing exercises, some serious problems with the crank-style cocking mechanism of the M3 occurred. Shortly thereafter, other units in Britain reported broken handles during their exercises in preparation of the imminent invasion of the European mainland. A subsequently arranged examination showed that the retracting pawl on the cocking handle was cracking adjacent to the rivet hole, where the pawl is riveted to the brace. An investigation revealed that the metal that was being used in the manufacture of the parts was not well suited to the heat-treating process. As a result, the parts became brittle and prone to breakage. Immediately the correct type of metal was substituted for the parts involved, and the width of the metal at the rivet hole was increased. Already-issued weapons had to be rebuilt with newly supplied parts by the armorer of the respective units.

A further improvement during production concerned the barrel ratchet pad; it had to be depressed to allow the barrel to be removed. But the early type was too small and too difficult to depress, and it was wearing prematurely. The new type was increased in size, and a change in the heat-treating process was instituted.

The last change in the design of the M3 concerned the rear peephole sight, in March 1944. The simple L-shaped sheet metal was not robust enough for the hard frontline use and was easily bent. The sight was made stronger by attaching lateral strengthening gussets.

All these improvements had been adopted in the ongoing production so that some hybrid forms of the M3 were created. However, the designation of the weapon did not change.

In May 1944, the Tank Destroyer Board submitted a number of additional suggestions:

- a longer stock
- a replaceable firing pin
- a spring clip or other means to attach the weapon to the shooter's belt
- more robust front and rear sights
- better material for the bolt face and the barrel chamber
- reinforced retracting pawl

An examination of the repeatedly occurring damages to the bolt face led to the conclusion that these were caused by the cleaning procedure. The soldiers used the rods of the stock as cleaning rods—but often with little care. The rod protruded through the barrel, and its end was rammed constantly against the bolt face. It had therefore been proposed to weld a transverse connection between the two rods to limit their insertion depth.

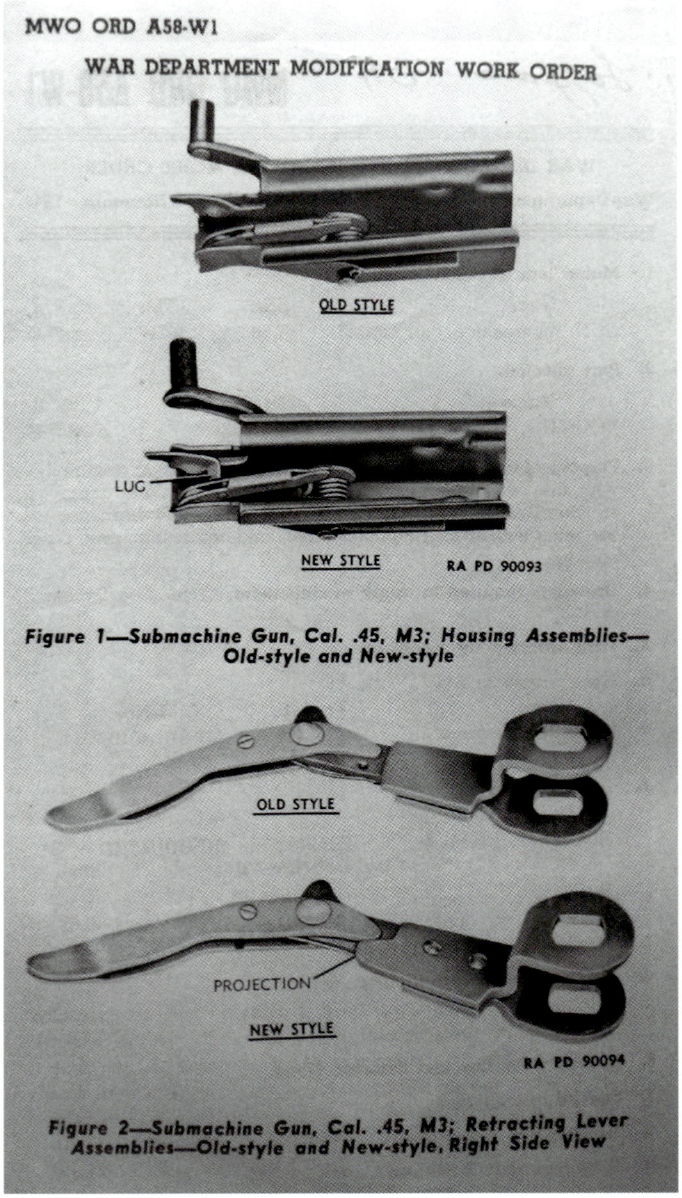

MWO ORD A58-W1

WAR DEPARTMENT MODIFICATION WORK ORDER

OLD STYLE

LUG

NEW STYLE

RA PD 90093

Figure 1—Submachine Gun, Cal. .45, M3; Housing Assemblies—Old-style and New-style

OLD STYLE

PROJECTION

NEW STYLE

RA PD 90094

Figure 2—Submachine Gun, Cal. .45, M3; Retracting Lever Assemblies—Old-style and New-style, Right Side View

Replacement of parts of the crank-style cocking mechanism (according to MWO ORD A58-W1). Weapons manufactured after serial number 173006 left the factory with the improved parts already installed.

Another flaw was the magazine release. There were numerous reports of the magazine catch being accidentally depressed, and the magazine falling out of the weapon. To avoid this problem, a small sheet-metal guard was installed around the magazine release button. The guard was held in place by a redesigned magazine release catch spring.

These changes to the stock and the magazine release were approved on August 31, 1944, and immediately integrated into the series production.

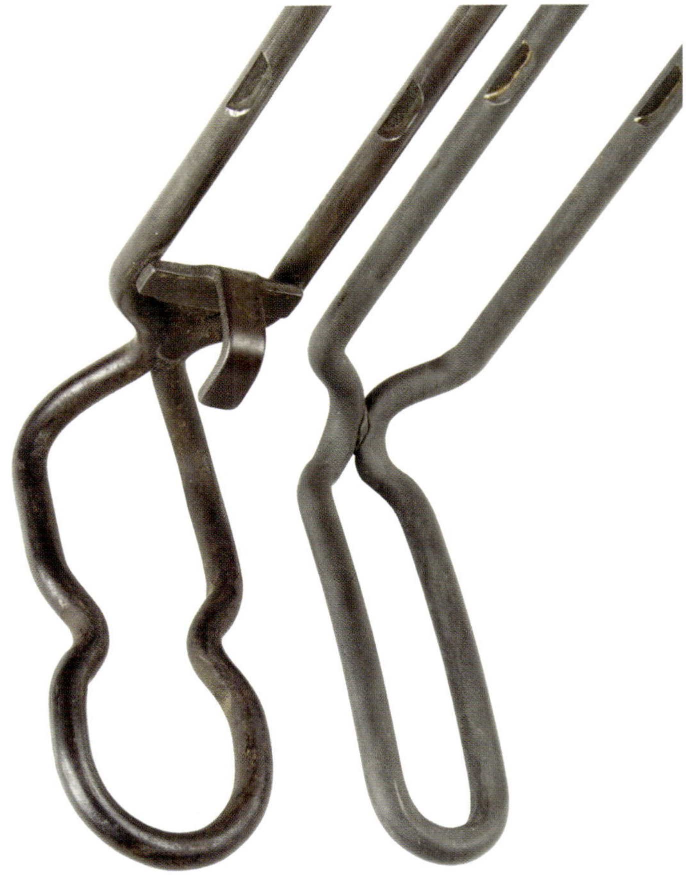

Comparison of the stocks: M3 (*right*) and M3A1 (*left*). Note the integrated magazine-loading tool.

M3 with reinforced rear sight

A sheet-metal guard (A7160997) and a redesigned magazine release catch spring (A7160998) were issued to protect the magazine release button (MWO ORD A58-W2/1945).

THE M3A1

Although the M3 was an extremely simple weapon to manufacture, the Ordnance Department still suspected potential for additional simplifications. A close examination revealed that the complicated crank-style cocking handle was a frequent source of failures, and that its elimination would also accelerate the series production. A way to simplify disassembly and maintenance was also under study.

On April 19, 1944, a formal development program was launched. Guide Lamp got the order for production of six submachine guns named "M3E1." The weapons were to include the following improvements:

- eliminate the crank-style cocking handle
- change bolt design to incorporate a finger hole for cocking purpose
- extend the ejection port cover rearward
- locate safety hook rearward on the ejection port cover to eliminate the safety contacting the bolt face
- locate the ejector in the receiver
- new magazine release button protective guard (already implemented in series production)
- new barrel ratchet assembly with an on/off barrel lock
- rear sight adjustable for manufacturing purposes
- relocate oiler in the hollow pistol grip
- install feed ramp on the guide rod locating plate
- new undercuts on the bolt
- new safety lock
- use of a conventional type of extractor

The listed changes were only suggestions, and not all of them were put into practice; some were found to be unsuitable or unnecessary. During the development of the M3E1, only the following changes had actually been implemented:

The six weapons of the M3E1 series (here the X33) were used to test improvements. Note the elongated shape of the ejector housing.

Left: Markings on the M3E1 with serial number X33
Above: Despite the elimination of the crank mechanism, the shape of the ejector housing remained unchanged on the M3A1.

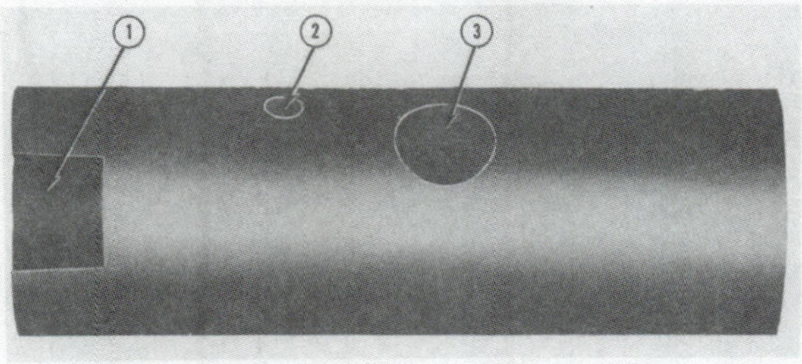

1. Clearance cut to allow ejection of spent cartridge case
2. Extractor pin
3. Safety lock recess

Figure 21. Top view of M3 Bolt.

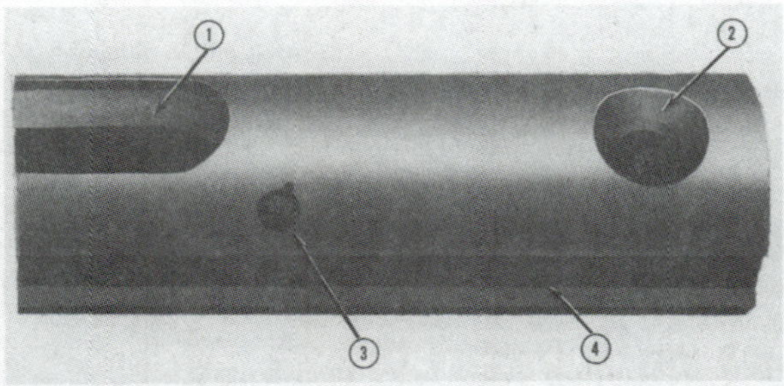

1. Cocking slot
2. Safety lock recess
3. Extractor pin
4. Clearance groove for cover rivets

Figure 22. Top view of M3A1 bolt.

Comparison of the bolts: M3 (*top*) and M3A1 (*bottom*). Note the finger hole (*1*) to cock the M3A1 (figures 21 and 22 from FM 23-41).

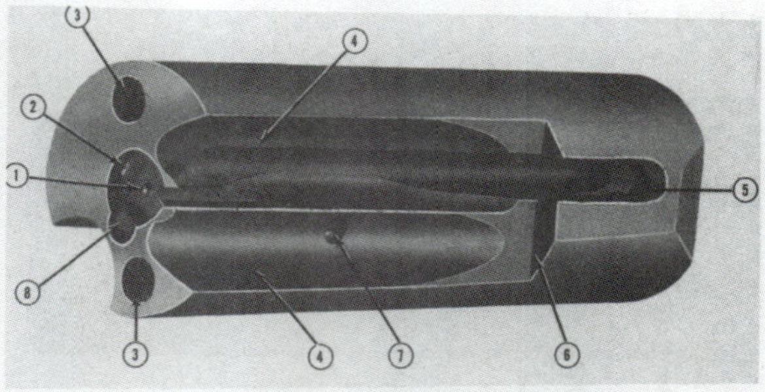

1. Firing pin
2. Cartridge recess
3. Guide rod bearing
4. Magazine lip recess
5. Retracting lever pawl notch
6. Sear notch
7. Extractor pin
8. Extractor

Figure 23. Bottom view of M3 bolt.

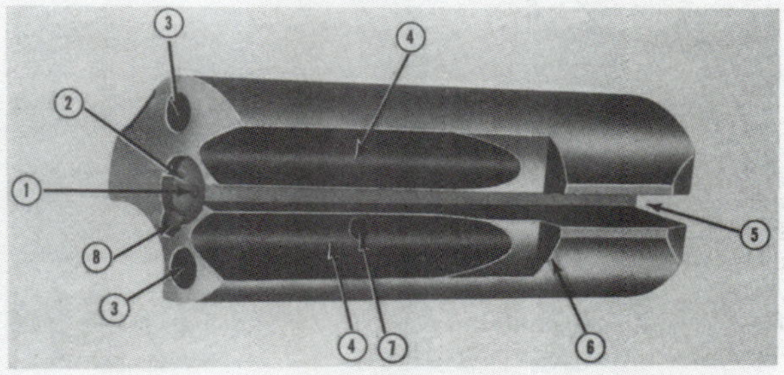

1. Firing pin
2. Cartridge recess
3. Guide rod bearings
4. Magazine lip recess
5. Ejector groove
6. Sear notch
7. Extractor pin
8. Extractor

Figure 24. Bottom view of M3A1 bolt.

Thanks to the full-length groove (*5*) for the ejector, the bolt of the M3A1 could be removed easily (figures 23 and 24 from FM 23-41).

Left: View from the front on the bolt assembly of the M3A1. *Courtesy of Mike Stannard*

Above: Comparison of the guide rod retaining plates: M3 (*left*) and M3A1 (*right*)

Cleaning brushes could be screwed into the threads in the stock ends.

The space inside the hollow pistol grip was used to install an oil can. The clip for the separate oiler was eliminated.

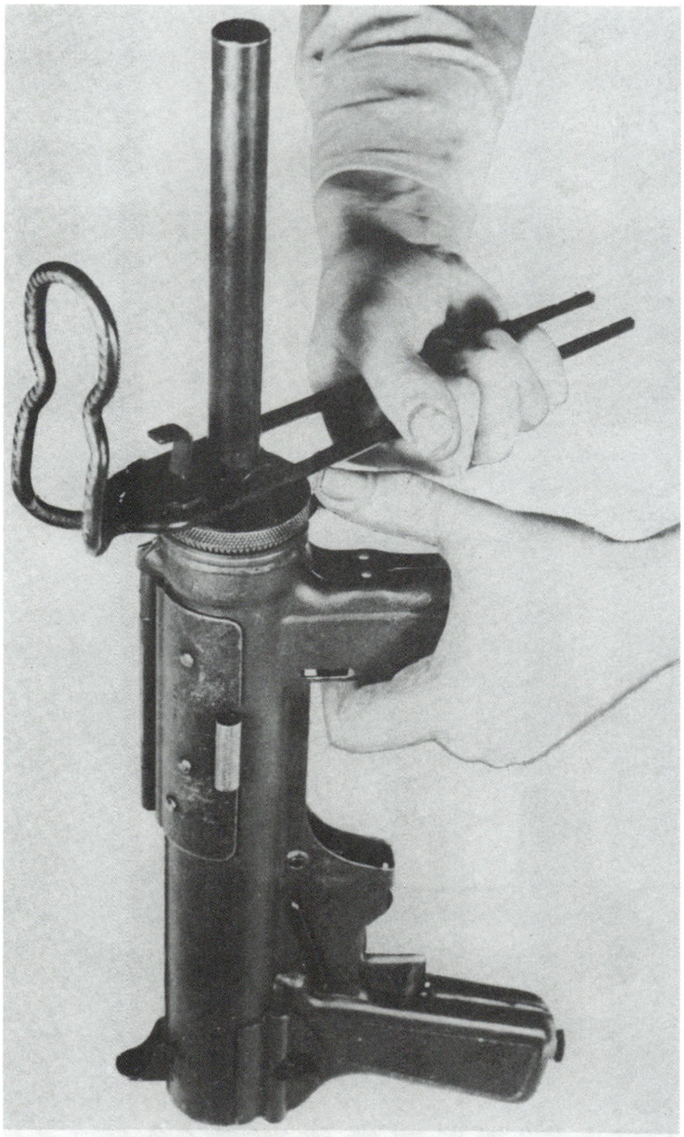

Figure 10. Using M3A1 stock as a wrench.

The barrel of the M3A1 got two wrench flats, so that the wire stock could be used as a removal tool (figure 10 from FM 23-41).

1. Elimination of the complete cocking assembly. Redesign of the bolt to be cocked by means of a finger hole. Enlargement of the ejection port and extension of its cover. Relocation of the cover's safety hook rearward. New hinge with stronger cover springs riveted to the receiver.
2. Cut of a groove for the ejector in the full length of the bolt, as well as a redesigned guide rod retaining plate to permit removal of the bolt assembly without removing the ejector housing.
3. Redesign of the stock, so that it could be used as a magazine-loading tool. The separate magazine filler was no longer needed. The new loading "hook" at the stock also served as a length limitation on insertion of the rods into the barrel.
4. The ends of both rods were drilled and tapped to accommodate a bore-cleaning brush (which had already been implemented in the current series production of the M3).
5. Incorporation of a larger oil can in the hollow pistol grip
6. Attaching two wrench flats to the barrel nut to allow its removal, using the wire stock as a tool
7. Protection of the magazine release button by a sheet-metal guard to avoid accidental release (which had already been implemented in the current series production of the M3)
8. Redesign of the L-type rear sight by attaching lateral strengthening gussets (which had already been implemented in the current series production of the M3)

Of the six finished model M3E1s, two guns were sent to the Infantry Board at Fort Benning, Georgia, and two others to Aberdeen Proving Ground. Since the modifications of the weapon had little effect on its functioning, the Ordnance Department decided to forego the repetition of the dust and mud tests. Only one shooting test was carried out in the presence of representatives of the US Army and the director of Infantry Boards. The reliability remained unchanged, and the durability of the incorporated improvements gave no cause for criticism.

An M3A1 (serial number 629058) with flash hider M9

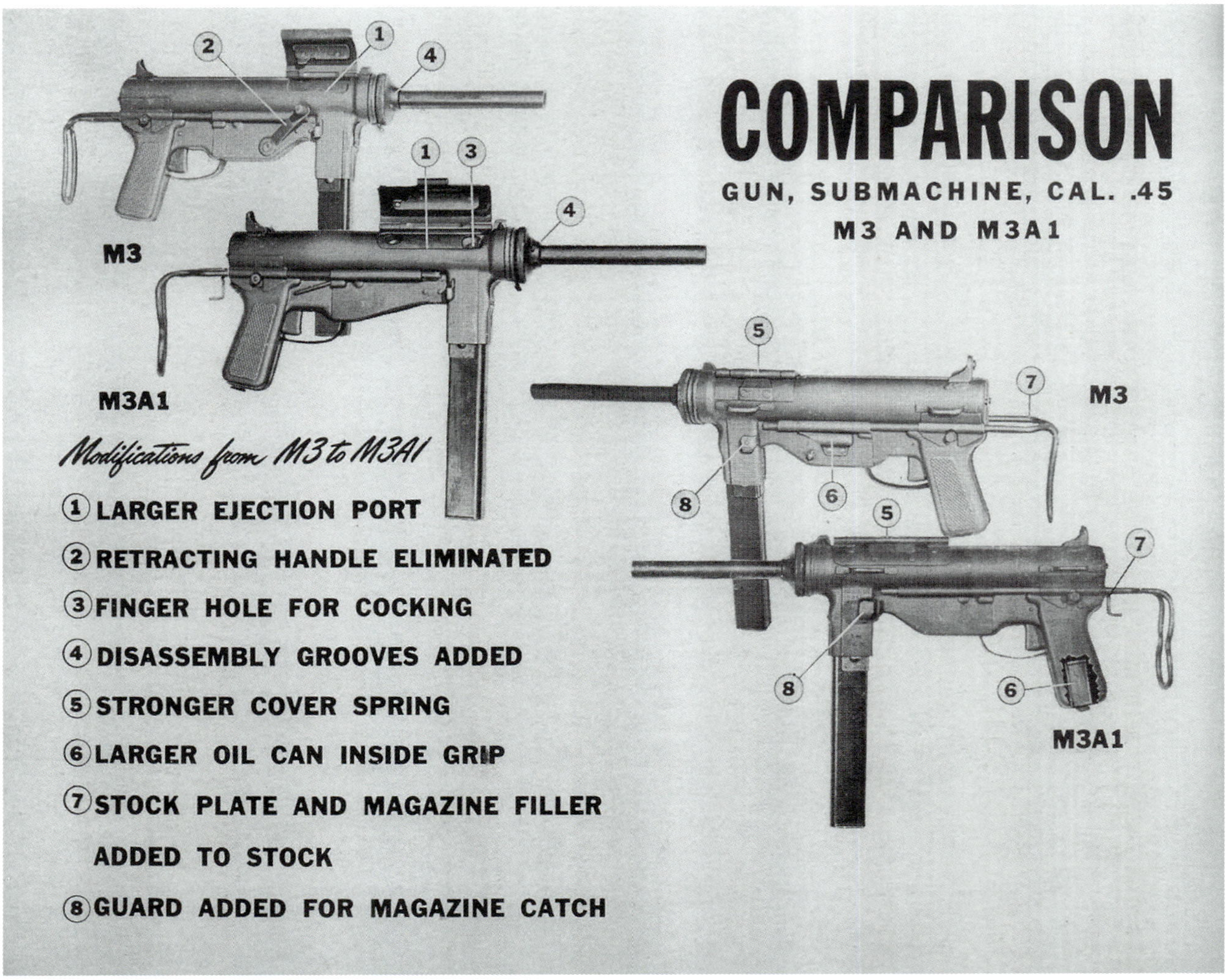

Educational material of the US Army, showing the most important differences between the M3 and M3A1

The adoption of the new model was recommended on November 2, 1944 (OCM 25593), and approved on November 21, 1944 (OCM 26148). Its new designation was "U.S. Submachine Gun, Caliber .45, M3A1."

Nothing else was changed with the basic design of the Grease Gun. However, during the preparation for use in Korea, in 1950, it was noticed that some weapons fired uncontrollably until the magazine was empty. An investigation found out that this was caused by unusually heavy wear of the bolt and sear of the M3. On the M3A1 the problem did not occur. The reason for the malfunction in the M3 was an incorrect heat treatment of these parts during the manufacturing process. So before sending the weapons to Korea, all the bolts had to be removed and treated again to achieve the required hardness. As a precaution, all sears were replaced by a new reinforced version.

During the Vietnam War, the Ithaca Gun Company in Ithaca, New York, made new barrels with chromed bores to achieve better protection against corrosion in the humid and warm atmosphere. Until the adoption of the M14 in 1957, the M3A1 remained the standard submachine gun of the US Army. The semiautomatic rifle M14 in 7.62 × 51 mm NATO caliber should replace the M1 Garand, the M1 carbine, and the submachine gun—in theory. But in combat, many soldiers continued to use the handy and reliable M3/M3A1. Even in the armored divisions of the National Guard, it was used until 1990.

FIELD MODIFICATIONS

The problems with the crank-style cocking handle that had already been determined in Britain during the preparations for the invasion could never be fully eliminated. The supply of improved exchange parts was not enough to upgrade all the weapons before Operation Overlord was launched. In France, many M3s became unusable because of failures in the cocking mechanism. Unfortunately, for a quick field repair without spare parts, the mechanism was too complicated. Therefore, some armorers converted the defective submachine guns to a "straight-pull cocking." They cut a longitudinal slot into the upper right side of the receiver and added a pin-style cocking handle to the bolt. In the forward position of the bolt, the cocking handle could be pulled out, so that removal of the bolt assembly from the receiver was still possible. The design is very similar to the British Sten submachine gun, which may have served as an inspiration for the modification.

Many weapons also got an improvised guard to protect the magazine release, to avoid accidental release of the magazine. Another improvement was attaching a limiter to the early-type stock, so that the rods would stop before hitting the bolt face when the barrel was being cleaned.

One of the modified M3s fell into German hands, but at that time the Germans presumably did not know that the M3 was usually cocked by a crank-style mechanism. Either way, the design did not make a great impression.

An interesting hybrid gun was discovered during the research for this book: an original standard M3 was heavily modified to create an M3A1. The crank-style cocking handle had been removed. A U-shaped piece of sheet metal was inserted into the now-empty ejector housing and fixed by screws through the side holes (in which previously the crank was installed). The bolt is from the M3A1. The ejection port was enlarged to the rear and closed by an M3A1 dustcover, which in turn was adapted to the existing M3 hinge.

It is unknown at what time and by whom this work was done. Unfortunately, at some point the weapon was deactivated, by cutting off the bolt face and welding the remaining rest immobile into the receiver. While doing this, the ejector was lost. The serial number was ground off, and the whole gun was painted black.

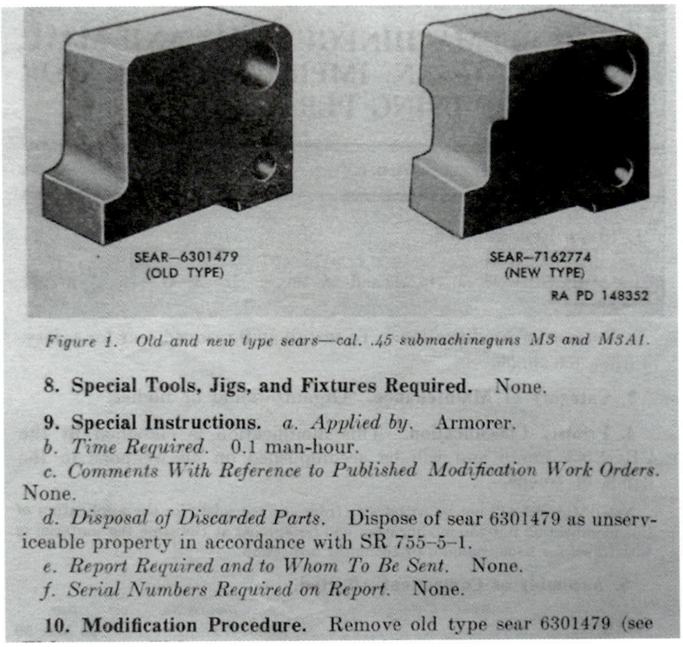

SEAR–6301479
(OLD TYPE)

SEAR–7162774
(NEW TYPE)

RA PD 148352

Figure 1. Old and new type sears—cal. .45 submachineguns M3 and M3A1.

8. Special Tools, Jigs, and Fixtures Required. None.

9. Special Instructions. *a. Applied by.* Armorer.
b. Time Required. 0.1 man-hour.
c. Comments With Reference to Published Modification Work Orders. None.
d. Disposal of Discarded Parts. Dispose of sear 6301479 as unserviceable property in accordance with SR 755–5–1.
e. Report Required and to Whom To Be Sent. None.
f. Serial Numbers Required on Report. None.

10. Modification Procedure. Remove old type sear 6301479 (see

Instruction for replacing the sear published in Modification Work Order (MWO ORD) A58-W3 from 1954

New chromed barrels were issued during the Vietnam War. *Courtesy of Frank Iannamico*

An M3 modified for cocking with a Sten-type cocking handle. The handle can be pulled out at the round opening at the front end of the slot. *Courtesy of the Musée National d'Histoire Militaire, Diekirch, Luxembourg; photo by Roland Gaul*

On this weapon, the no-longer-used lateral openings in the ejector housing were closed. *Courtesy of the Musée National d'Histoire Militaire, Diekirch, Luxembourg; photo by Roland Gaul*

The cocking handle was not permanently connected to the bolt. Note the lateral grooves to slide along the edges of the receiver slot. *Courtesy of the Musée National d'Histoire Militaire, Diekirch, Luxembourg; photo by Roland Gaul*

This rim was added to protect the magazine release button from damage and to avoid accidental release. *Courtesy of the Musée National d'Histoire Militaire, Diekirch, Luxembourg; photo by Roland Gaul*

A different type of self-made guard to protect the magazine release button

Soldiers of the 95th Infantry Division at the border of the German Reich. Note the M3 with modified cocking handle. The soldier on the right carries an SS dagger as a souvenir.

A modified M3 in use with the 739th Medium Tank Battalion. Here a Sherman M4 equipped with the T1E3/M1 Mine Exploder gets recovered.

Relic of a modified M3 found on a beach in Normandy. Since the receiver is stripped of nearly all other parts, it can be assumed that it was at some point disposed of into the sea.

Field modification of an M3 stock with limiters to avoid damage to the bolt face when cleaning the barrel. *Courtesy of Ferdinand Weiss*

Transformation of an M3 into an M3A1. The serial number was ground off.

The enlarged ejection port with adapted M3A1 dustcover

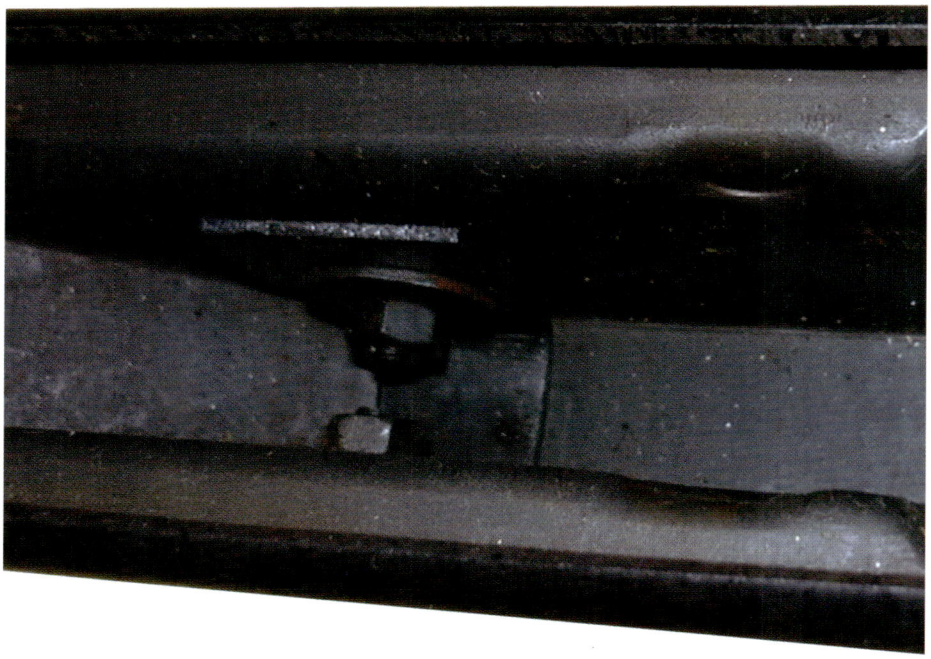

The lateral openings in the ejector housing were closed by adding a U-shaped piece of sheet metal.

07 MANUFACTURERS AND PRODUCTION

GUIDE LAMP DIVISION / GENERAL MOTORS CORPORATION, ANDERSON, INDIANA

The award of the design contract to Guide Lamp was well considered. The company, founded by William F. Persons and William Bunce in Cleveland, Ohio, in 1906, had longtime experience in sheet-metal processing. After some initial difficulties, the production of vehicle headlights for the burgeoning automobile industry proved to be a lucrative business. In the following years, the Guide Lamp Manufacturing Corporation became the market leader, supplying all major vehicle manufacturers in the United States. The number of employees reached 496 in the year 1927. At that time, General Motors was striving for its own production of headlights and thus acquired Guide Lamp in August 1928, to integrate the company into their Delco-Remy Division. A year later, Guide Lamp became a separate division with two factories: Cleveland, Ohio, and Anderson, Indiana. The number of employees reached 2,374 in 1936. At the outbreak of World War II, the demand not only for military vehicle headlights but also for other military equipment was increasing. So, in 1942, Guide Lamp began with the additional production of aircraft parts, engines, and weapons.

Until the end of the war, the following equipment rolled off the production lines:

- 36,750,000 brass and steel cartridge cases in five different sizes
- 8,500,000 headlights and lamps of all kinds
- 3,400,000 plastic "Stimsonite" reflectors for vehicles
- 1,600,000 barrels for the Browning machine gun M2 (daily output of 2,000 pieces)
- 1,000,000 water-jacket sleeves for Allison aircraft engines
- 22,000 spinner noses and adapters for the Bell P-39 Airacobra

For this achievement the company received the desired Army-Navy "E" Award with additional four stars.

In 1943, 38 percent of the company was making headlamps, 29 percent sleeves, and 17 percent machine gun barrels, with the remaining 16 percent busy with the M3 Grease Gun. At that time, about a third of the workforce consisted of women, since the company had already lost more than 1,000 workers for military service.

At the end of 1944, the number of employees reached its peak at over 5,000. By 1946, only 3,645 people were still employed. In the postwar period, Guide Lamp continued its successful operation in the market of automobile headlights. But the growing competition from foreign countries and the general recession in the automotive industry in the 1980s made life hard for General Motors. In 1984, Guide Lamp and Fisher Body united to form "Fisher Guide," which was renamed "Inland Fisher Guide" in 1990. Persistent losses and gloomy forecasts led to a series of job cuts and reemployment programs in the General Motors group. Ultimately, the company was sold off by General Motors in 1998 and became the independent Guide Corporation. But the new management could not improve the economic situation. In view of a loss of 200 million dollars in 2004, and an expected deficit of 260 million dollars in 2006, the end was near: the Guide Corporation announced the closure of their

Aerial view of the Guide Lamp plant in Anderson, Indiana, during the war. At the best times, more than 5,000 people were employed there.

plant for mid-2007. But on January 12, 2007, the last shift finished their work, six months earlier than planned. Nearly 1,400 workers became unemployed overnight. The factory buildings remained for two years before they were demolished in 2009–10. Even today, no new buildings have been erected there. The contaminated soil must be disposed of—but no investor is willing to spend money on it.

When the contract for production of 300,000 submachine guns was issued to Guide Lamp in January 1943, the agreed-on price for one weapon without bolt was $17.92. Later, the price had to be raised to $18.36 due to small design changes and increased costs of packaging and shipping. The bolts were not manufactured by Guide Lamp but were delivered by the subcontractor Buffalo Arms Company at a price of $2.58. A complete Grease Gun cost only half of a Thompson M1A1, due to the reduced production time and material requirements.

The M3 was primarily made from simple .060-inch-thick sheet metal, which was formed into shape by die-stamping. Only two major parts, the barrel and the bolt, required machining. The receiver was welded together lengthwise from two sheet-metal halves. Ironically, this simple-sounding operation caused more trouble than expected: the thin metal of the receiver halves was warping from the heat of the welding process. This problem did not occur when the T-20 guns were carefully manufactured for the

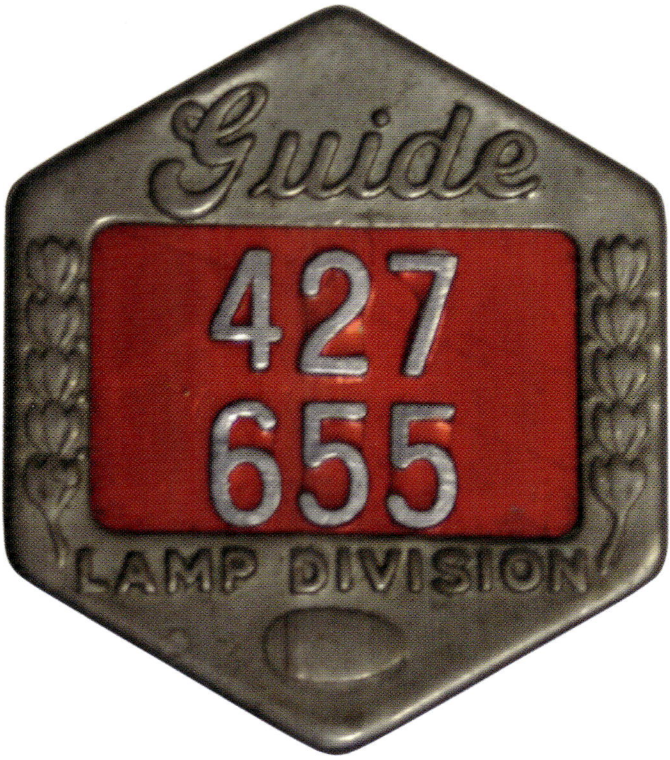

Badge of an employee of Guide Lamp. *Courtesy of David R. Albert*

In September 1942, Guide Lamp was awarded the Army-Navy "E" Award, followed by four additional stars until the end of the war. The award flag is held by Lt. Col. W. R. Martin (executive officer of the Cincinnati Ordnance District) and C. A. Michel (general manager of Guide Lamp).

The employees got a certificate and pins.

testing, but now, in the large-scale series production, things were different. The planned output of 20,000 weapons in July 1943 could not be reached. Only 900 weapons passed the final quality control. The problems with the receiver halves were so severe that the Ordnance Department had to sign a contract with Auto-Ordnance about the delivery of additional Thompson submachine guns to fill the gap. Only after a few weeks later did Guide Lamp get things under control, and the series production could run as planned.

To save valuable working hours, the barrels were no longer traditionally made. A method called "cold swaging" was used instead: the steel tubing for the barrel was first reamed to the correct inside diameter. Then the rifling was pressed into the barrel by inserting a mandrel and compressing the tube as it was forced through a die. After this process, the barrel was heated so that the mandrel could be removed more easily. The 40-inch piece of tubing was then cut in sections to produce five M3 barrels. The accuracy of fire remained well within the requirements set up by the Ordnance Department. The British also used this method for making their Sten barrels.

After assembly of all parts, the weapons were handed over to ready the sight. This interesting process took place in a long row of shooting booths and is described in a Guide Lamp brochure: "The front sight is affixed in the assembly as is the blank for the rear sight. However, the rear peep is not drilled until the gun is tested for targeting. Then, by an ingenious telescopic sight developed by Guide engineers, the peep is drilled with direct reference to five shots fired from the gun. When the telescope cross-hairs are trained on the bulls-eye pattern, the exact proper location of the rear peep is indicated by a drill guide. The drilling then is done before the gun is moved. After drilling, the sighting is tested by another burst of fire without the use of the telescope." Subsequent adjustment was not possible.

Promotion booklet published by Guide Lamp in 1944

The M3 is also mentioned in the booklet: "Soldiers used it for close-in fighting—like in the jungles of the South Pacific—and in the rocky, rugged hills of Italy. It was also a favorite weapon for street fighting in the Italian campaign. Mostly, it was fired from the hip—pouring a murderous stream of .45-caliber slugs in short, deadly bursts. It became extra popular because it could be held on the target so easily—didn't 'climb' as had earlier submachine guns."

Engineering draftsmen at work

One of the many workshops at Guide Lamp

Guide Lamp had its own toolmaking shop.

The Guide Lamp Girls Bowling Team in 1940–41. In the course of war more and more women were employed by the armament industry.

Markings on a bolt made by Buffalo Arms in June 1944. *Courtesy of Rob Stott*

Pressing of the left receiver half

Punching out the right receiver half

Left: Welding together both receiver halves was not an easy task. Note the tally sheet on each workstation to count the finished receivers.

Right: Adjusting the barrel blanks before further processing

The rifling was pressed into the barrel by inserting a mandrel and compressing the tube as it was forced through a die.

One blank was cut in sections to produce five M3 barrels.

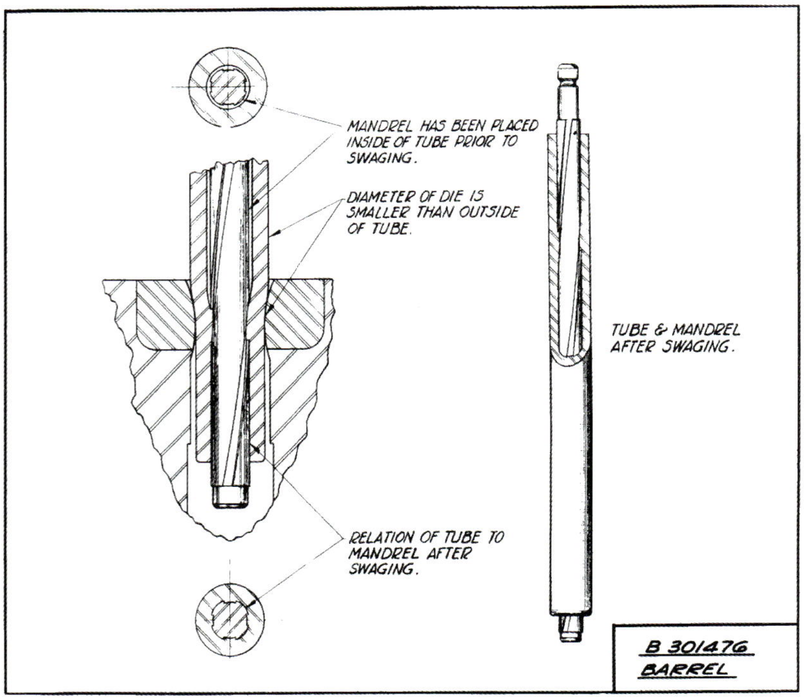

Illustration of the "cold swaging" method in the Guide Lamp production documents compilation

The M3 is taking shape: in the foreground the extractor is added to the bolt, and in the rear all the individual components are put together to finish the M3. Note the high proportion of women.

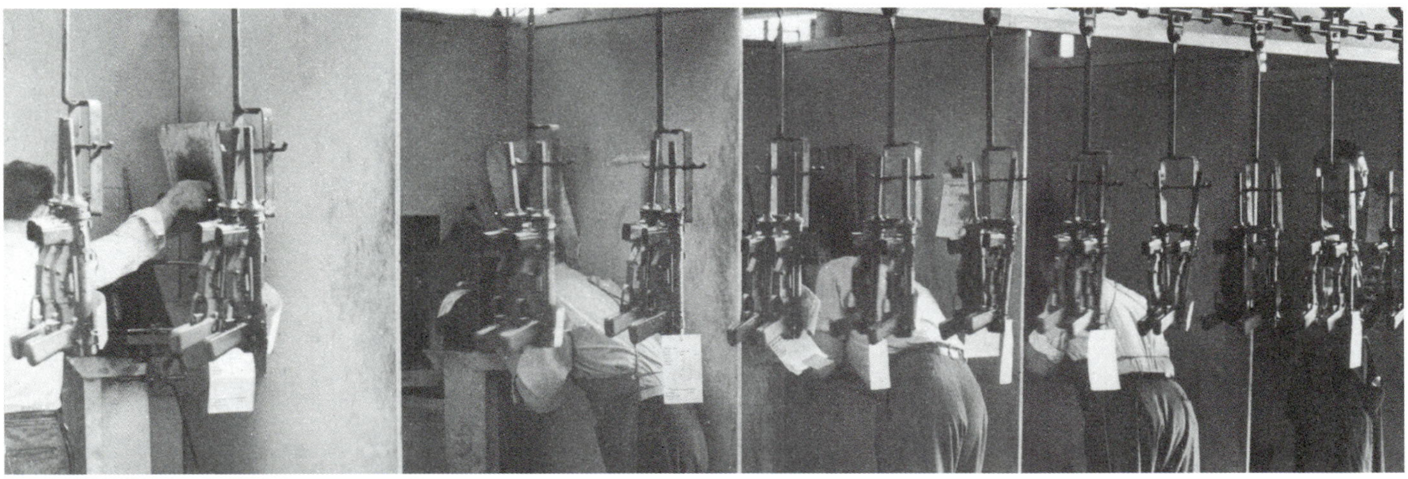

The row of shooting booths for adjusting the rear sight. The oblique boards were installed to deflect the spent cases downward.

After adjustment, the peephole was drilled into the rear sight.

Comparison of two drilled peepholes

A special magazine loader saved valuable working time.

The adjusting and testing procedures of the weapons consumed a lot of ammunition. Note the transportation system with M3s running below the ceiling.

Then the weapons were thoroughly cleaned and phosphated before they were handed over for the final inspection. This final step was not documented with an additional stamp on the weapons, as was common practice on weapons in other countries. In June 1944, an average of 1,000 M3s were made in twenty-four hours.

Fixing of an M3 into a mount for test shooting

SUBCONTRACTORS OF GUIDE LAMP, 1942–45

For reasons of capacity, Guide Lamp could not make all parts of the M3/M3A1 on its own. So, contracts for individual parts were awarded to more than twenty subcontractors.

To keep the quality of the delivered weapons at a constantly high level, individual weapons were randomly taken from the current production and extensively tested by the Ordnance Department:

1. durability:
   a. Weapons and spare barrels are required to withstand the firing of one high-pressure proof cartridge.
   b. Weapon has to pass the 5,000-round endurance test.
2. handling: Weapons shall function without problems. Rate of fire not more than 450 rounds per minute.
3. precision: Four out of five shots must hit a 5-inch bull's-eye at a range of 100 feet.
4. marking: All gun parts shall be clearly marked according with the applicable drawings.
5. workmanship: Shall be in accordance with the requirements of MIL-W-13855.
6. packaging: The packaging must comply with the packaging guidelines.

Final inspection for defects and visible damage. Every individual work step has been noted on the consignment slip and signed by the respective performing employee.

Loading of boxes with new M3s onto a truck

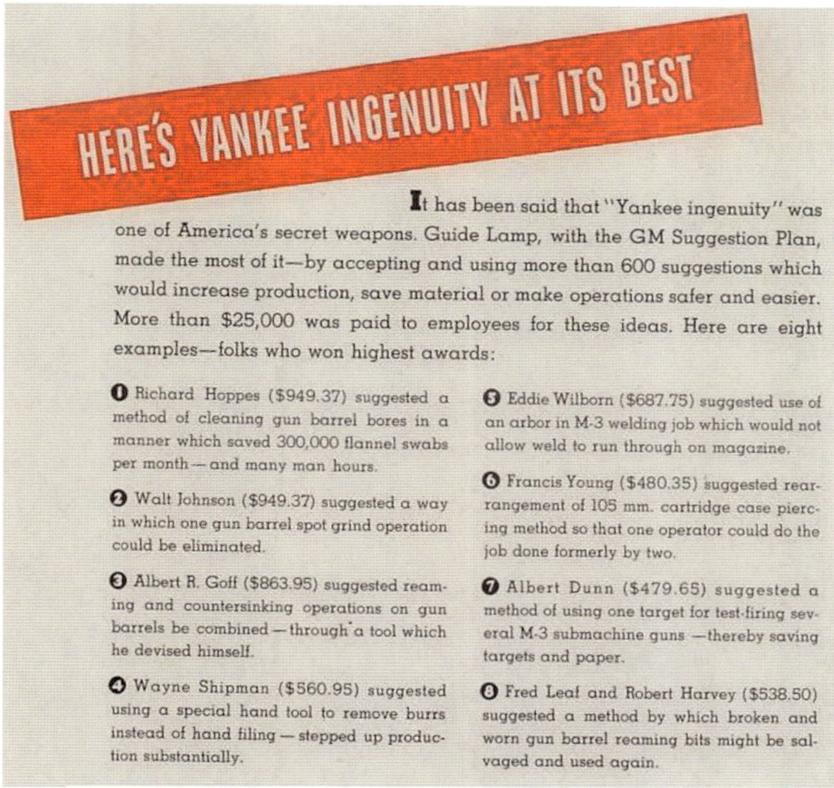

HERE'S YANKEE INGENUITY AT ITS BEST

It has been said that "Yankee ingenuity" was one of America's secret weapons. Guide Lamp, with the GM Suggestion Plan, made the most of it—by accepting and using more than 600 suggestions which would increase production, save material or make operations safer and easier. More than $25,000 was paid to employees for these ideas. Here are eight examples—folks who won highest awards:

❶ Richard Hoppes ($949.37) suggested a method of cleaning gun barrel bores in a manner which saved 300,000 flannel swabs per month—and many man hours.

❷ Walt Johnson ($949.37) suggested a way in which one gun barrel spot grind operation could be eliminated.

❸ Albert R. Goff ($863.95) suggested reaming and countersinking operations on gun barrels be combined—through a tool which he devised himself.

❹ Wayne Shipman ($560.95) suggested using a special hand tool to remove burrs instead of hand filing—stepped up production substantially.

❺ Eddie Wilborn ($687.75) suggested use of an arbor in M-3 welding job which would not allow weld to run through on magazine.

❻ Francis Young ($480.35) suggested rearrangement of 105 mm. cartridge case piercing method so that one operator could do the job done formerly by two.

❼ Albert Dunn ($479.65) suggested a method of using one target for test-firing several M-3 submachine guns—thereby saving targets and paper.

❽ Fred Leaf and Robert Harvey ($538.50) suggested a method by which broken and worn gun barrel reaming bits might be salvaged and used again.

Clever employees could get awards for suggestions for improvements. The best suggestions were published to encourage other employees.

Lt. C. L. Asmann, Army inspector of ordnance, inspects an M3 taken from the current series production.

Markings on an early M3 (serial number 14133) reworked by Rock Island Arsenal (RIA) between 1947 and 1958. "EB" is the personal stamp of ordnance inspection supervisor Elmer M. Bjerke.

Markings on a late M3 (serial number 496553). The number is now positioned below the US Army Ordnance Corps symbol.

Serial number with corrected last digit (240241) of an M3

Markings on an M3A1 (serial number 629058)

Murray Van Wagoner (then governor of Michigan) is shooting a machine gun at the Saginaw company. Behind him, Saginaw manager Alva Phelps can be seen. *Courtesy of the General Motors Heritage Center, Sterling Heights*

Western Automatic, founded in 1882, was the oldest company in Elyria.

Middle: During testing in Aberdeen, the welded connection between the barrel bushing and receiver broke on this M3 (August 1943). *APG*

Bottom: Members of the Women's Army Corps (WAC) are test-shooting the M3s.

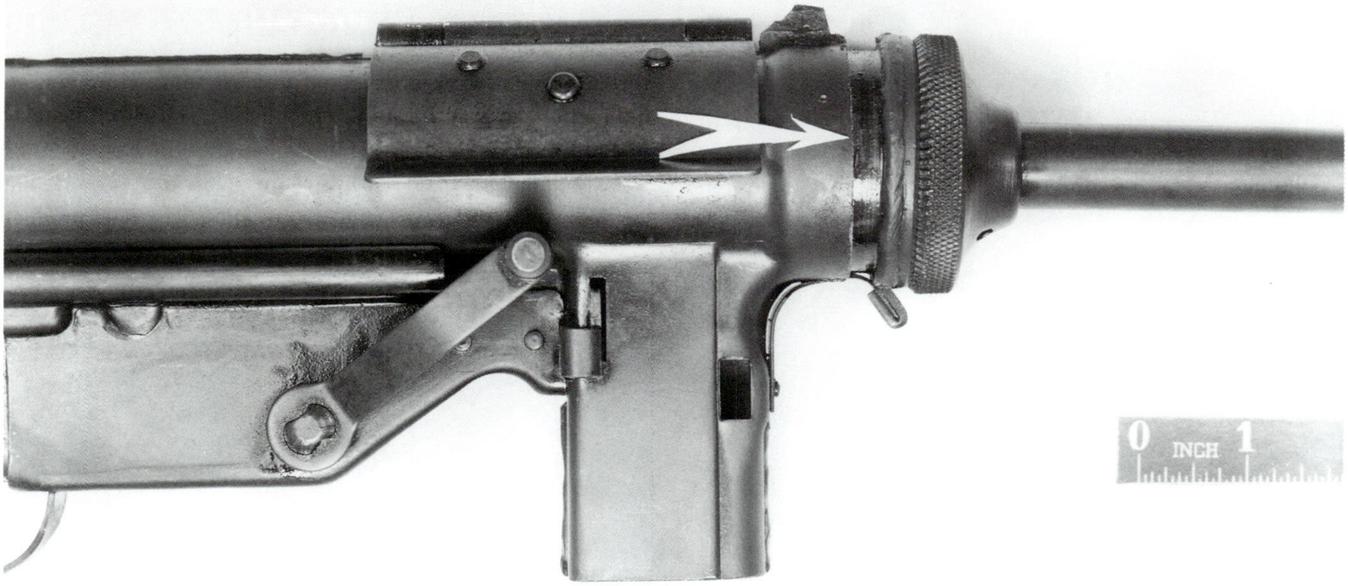

The total output of Guide Lamp from 1943 to August 1945 was 688,975 Grease Guns:

1943	85,130 M3s
1944	343,372 M3s
1945	178,192 M3s and 82,281 M3A1s

This amount corresponds to nearly one-third of all manufactured Thompsons in World War II.

On February 22, 1945, the Defense Department had decided that the Grease Gun (M3 and M3A1) should replace the Thompson models 1928, M1, and M1A1 (OCR 26747). The objective was to reduce the variety of models and thus result in a streamlined supply of spare parts and maintenance work. Due to the high production figures, a rapid exchange of the weapons was assumed. But after the end of the war in Europe and the looming defeat of Japan, the production was ceased in mid-1945. The last M3A1 rolled off the line in July 1945:

April	30,117 M3A1s
May	26,000 M3A1s
June	17,164 M3A1s
July	9,000 M3A1s

However, it was not possible to fully remove the Thompson models from active service. According to a report of the US Army from March 27, 1945, there was a lack of about 50,000 Grease Guns. Only after the acquisition of weapons from Ithaca production after the Korean War was the active duty of the Thompson officially ended on March 14, 1957.

Production Documents from Guide Lamp

As already described, Guide Lamp had to overcome some initial difficulties in the manufacture of the M3. And even then, there have been minor improvements in the manufacturing or assembly of parts. All these lessons learned were summarized by Guide Lamp in June 1944, in the ***Description of Manufacture / Gun, Submachine, Caliber .45, M3***.

The compilation contains 400 pages on the exact description of the manufacturing of every part, with equipment lists, labor requirements, technical drawings, and images of the respective steps. The letter before each drawing number stands for the size of the sheet according to the Ordnance Department standard: A = 8.5 by 11 inches, B = 11 by 17 inches, C = 17 by 22 inches, D = 22 by 34 inches.

ITHACA GUN COMPANY, ITHACA, NEW YORK

When it became clear that the Korean War (June 1950 to July 1953) would last longer than expected, the US Army began to check their stored submachine guns (M3 and M3A1) to make them ready for deployment. In 1953, the Ithaca Gun Company in Ithaca received an order to manufacture spare bolts (contract number DA 30-115-ORD-355). In the two years from 1953 to 1954, about 90,136 complete bolts were delivered. Other manufacturers such as Springfield Armory and Rock Island Armory provided various other parts.

The Ithaca Gun Company was founded in 1880 by Leroy Smith and William Henry Baker. As a location for the factory building, they chose a place right on Fall Creek to use the power of water for energy generation. In subsequent years, the company reached a high level of awareness and reputation through its world-class hunting rifles and shotguns. In mid-1942, Ithaca Gun received its first military order: a license production of Colt's automatic pistol M1911A1. Already in December, the series production began, initially supported by parts supplied by Colt, Springfield Armory, Remington, and other manufacturers. Until the end of the war, Ithaca Gun produced 335,466 pistols.

In 1953, the Ithaca Gun Company was awarded a major contract by the Rochester New York Ordnance District: 70,000 complete submachine guns M3A1 (contract number DA 19-058-ORD-7894). When setting up the production facility and organization of the series production processes at Ithaca, the company counted on the extensive experience of Guide Lamp. Thanks to this support, the series production was expected to start soon. However, some difficulties delayed the start of production, so that ultimately the first M3A1 rolled off the line in 1955—after the Korean War had already ended. Large quantities of submachine guns were no longer needed, so the order was canceled in 1956. The total production

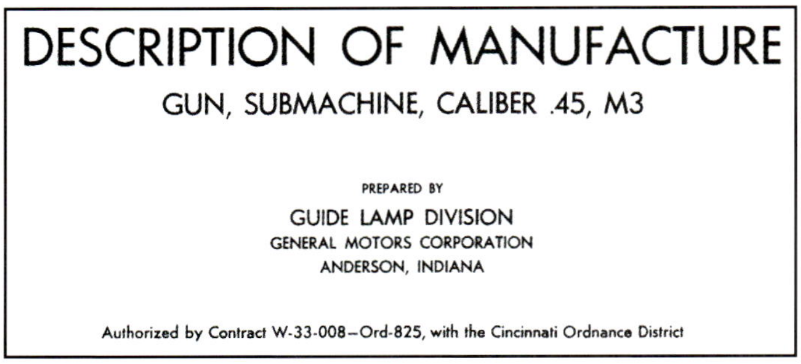
DESCRIPTION OF MANUFACTURE
GUN, SUBMACHINE, CALIBER .45, M3
PREPARED BY
GUIDE LAMP DIVISION
GENERAL MOTORS CORPORATION
ANDERSON, INDIANA
Authorized by Contract W-33-008–Ord-825, with the Cincinnati Ordnance District

The compilation *Description of Manufacture / Gun, Submachine, Caliber .45, M3* was distributed by Guide Lamp to subcontractors in June 1944. *Courtesy of the Rock Island Arsenal Museum*

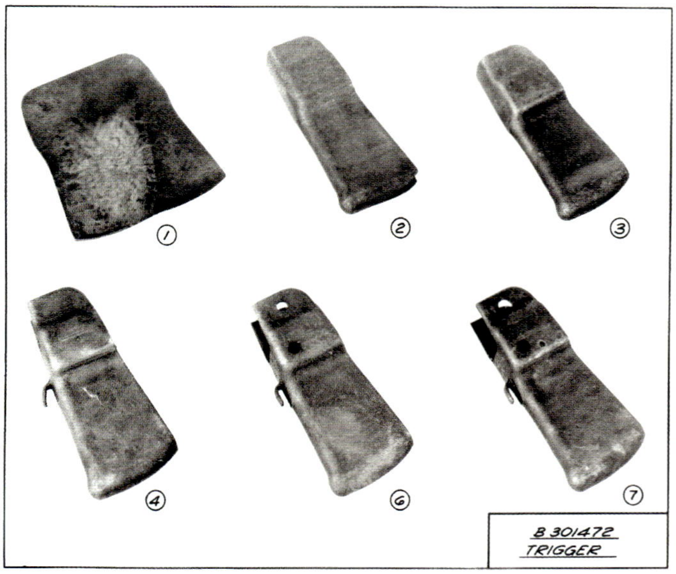

Illustrations of the seven working steps for making the trigger. *Courtesy of the Rock Island Arsenal Museum*

was only 33,227 weapons, which was little more than a single month's output of Guide Lamp during World War II.

The serial numbers of the weapons made by Ithaca directly continued after the last serial number of Guide Lamp. This ensured that no new weapon could get an already existing serial number. To date, serial numbers are known to range from 721330 to 754556. The M3A1s from Ithaca can be distinguished from the Guide Lamp ones only by the markings and a slightly different pattern of the pistol grip. All parts are interchangeable with those from wartime production.

During the Vietnam War, Ithaca made new barrels with a chromed bore to protect them against corrosion.

In 1967, the Smith family sold the Ithaca Gun Company to a company in Colorado, from which a short time later General Recreation Inc. originated. However, this company got into financial problems in the late 1970s and sold all subsidiaries—except for the Ithaca Gun Company. After an unsuccessful attempt to relocate the weapons manufacturing to Colorado, General Recreation Inc. had to declare bankruptcy in December 1978. This was followed by a few months of downtime, but then production was resumed. Unfortunately, the company went down a second time in 1985 and was then up for sale. The new owner (Ithaca Acquisition Inc.) moved the company headquarters to King Ferry, New York, in 1987. In 2005, the company received an investment loan of $150,000 from Cayuga County and moved to larger premises in Auburn, New York, but the organization of a functioning production at the new site failed for unknown reasons. This was followed by the sale to the Marshall family in Upper Sandusky, Ohio. There, numerous designs were revised and adapted for modern production on CNC machines. But once again, the business failed due to a lack of funds. In June 2007, Dave Dlubak acquired the company. Since the location remained unchanged this time, many employees could remain in the company. Today the Ithaca Gun Company successfully manufactures high-quality hunting and sporting weapons again.

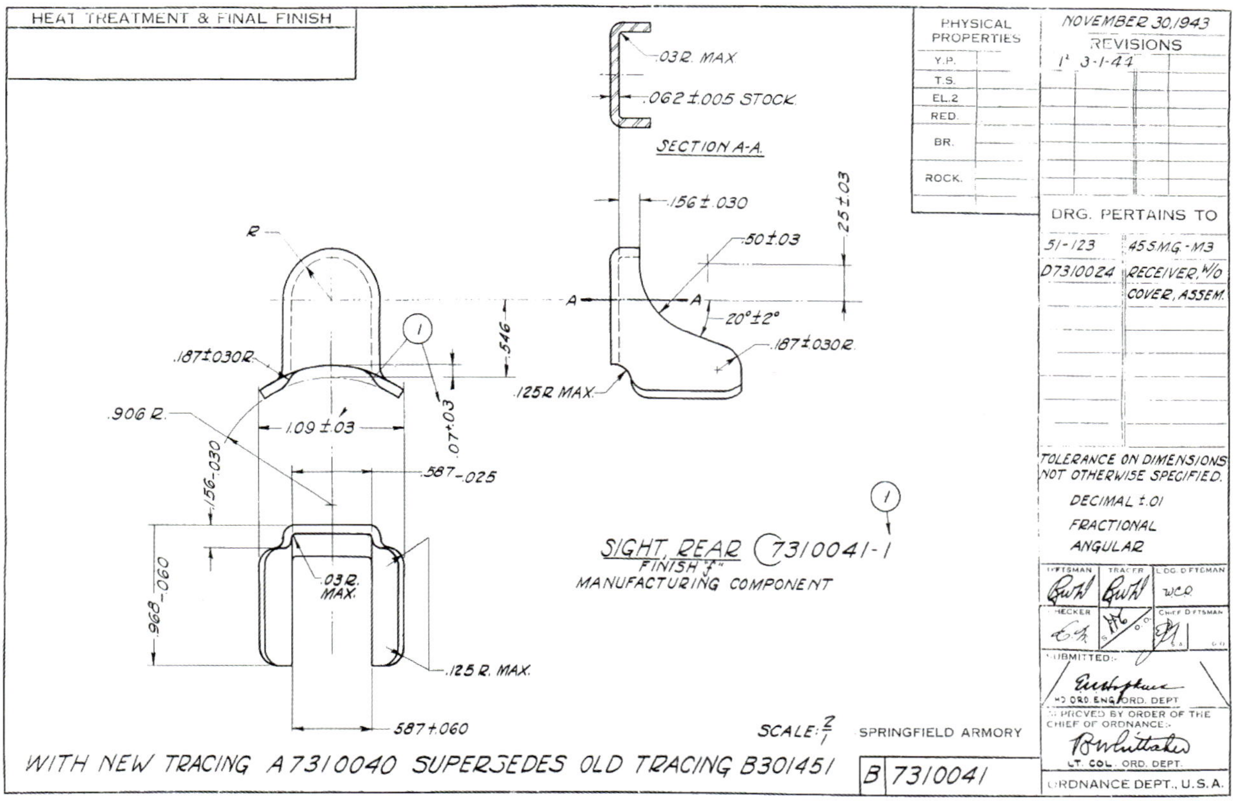

Also some design changes of individual parts are already included in the compilation, such as the reinforced rear sight. *Courtesy of the Rock Island Arsenal Museum*

GUIDE LAMP DIVISION
GENERAL MOTORS CORPORATION
ANDERSON, INDIANA

LIST OF MANUFACTURING OPERATIONS

FACIFACILITIES PRODUCING Guide Lamp Division PART NAME Trigger PART NUMBER B 301472

SEQ.	OPERATIONS DESCRIPTION	MAN HRS. PER 100 PCS.	MACH. HRS. PER 100 PCS.	MACHINERY & EQUIPMENT DESCRIPTION	FIXTURES, TOOLS & MFG. GAGES DESCRIPTION	TOOL NUMBER
1	Blank (1) at a time	.0834	.0834	#5 Minster Inclinable Press	Blanking Die	16043-A
2	Draw	.2434	.2434	#4-R Rockford Inclinable Press	Draw Die	16043-B
3	Restrike 1.788" radius	.2299	.2299	#4 Toledo Inclinable Press	Restrike Die	16043-D
4	Notch shear, and first form spring tang	.2271	.2271	#4 Toledo Inclinable Press	Notch and Form Die	16043-E
5	Restrike to straighten sides	.3209	.3209	Style "E" Cleveland Inclinable Press	Restrike Die	16043-K
6	Finish form spring tang and pierce (4) holes	.2218	.2218	#23 Consolidated Horning Press	Form and Pierce, Cam Die	16043-G
7	Open .129" diameter hole to .159"	.1633	.1633	#81-H Consolidated Bench Press	Pierce Die	16043-J
8	Restrike sides	.3108	.3108	#39 Bliss Horning Press	Restrike Die	16043-K
9	Remove burrs from inside	1.3889		Bench		
10	Degrease	.0070		Hand Tank		
11	Tumble	.0075		Globe Tumbling Barrel		
12	INSPECT				See Inspection Manual	

PAGE 106

List of working steps, man- and machine-hours, and kind of machines needed for making the trigger. *Courtesy of the Rock Island Arsenal Museum*

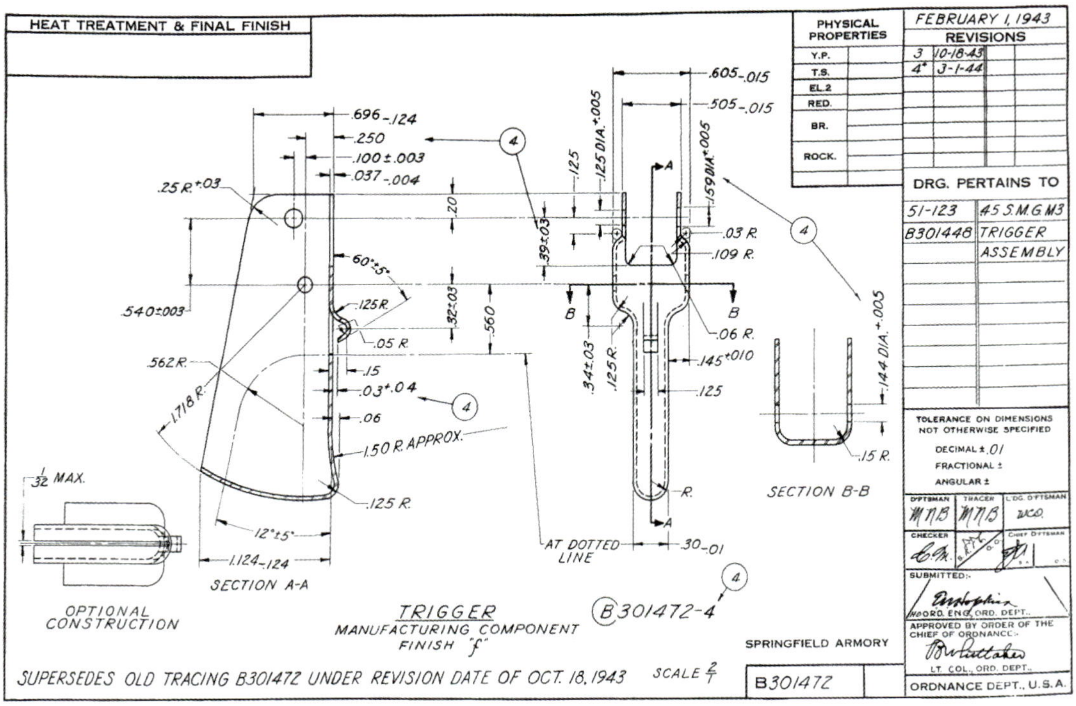

The compilation was continually updated. Here, for example, drawing B30/472 of the trigger from March 1, 1944, replaced the one from October 18, 1943. *Courtesy of the Rock Island Arsenal Museum*

GUIDE LAMP DIVISION
GENERAL MOTORS CORPORATION
ANDERSON, INDIANA

INSPECTION OPERATIONS FOR GUN, SUBMACHINE, CALIBER .45, M3

FACILITIES PRODUCING GUIDE LAMP DIVISION of General Motors Corporation. PART NAME TRIGGER PART NUMBER B-301472

APPLICABLE SPECIFICATIONS

MATERIAL U. S. ARMY SPECIFICATION #57-136B CLASS Steel WD-0
HEAT TREATMENT None PHYSICALS None
FINISH None (At Assembly)

SEQUENCE	DESCRIPTION OF INSPECTION OPERATION	GAGE DWG. No.	MAN HOURS PER 100 PIECES
1.	Receive stock and store.		
2.	Take sample of steel and make chemical analysis.		
3.	Blank.		
	FLOOR INSPECTION:		
4.	1% check .037 - .004 Dimension. Hand Micrometers	Micrometers	
5.	5% check visual for burrs and tool marks.	Visual	
6.	Draw		
	FLOOR INSPECTION:		
7.	5% check visual for burrs and tool marks.	Visual	
8.	Re-strike.		
	FLOOR INSPECTION:		
9.	5% check visual for burrs and tool marks.	Visual	
10.	Notch, shear and first form spring tang.		
	FLOOR INSPECTION:		
11.	5% check visual for burrs and tool marks.	Visual	
12.	Re-strike to straighten sides.		
	FLOOR INSPECTION:		
13.	5% check visual for burrs and tool marks.	Visual	
14.	Finish form spring tang and pierce four holes.		
	FLOOR INSPECTION:		
15.	5% check visual for burrs and tool marks.	Visual	

SHEET No. 1 OF 2 PAGE 158

Guidance for quality control during production of the trigger. The percentage number of parts to be checked and respective procedures were mandatory: chemical tests of materials and visual examination on burrs and tool marks. *Courtesy of the Rock Island Arsenal Museum*

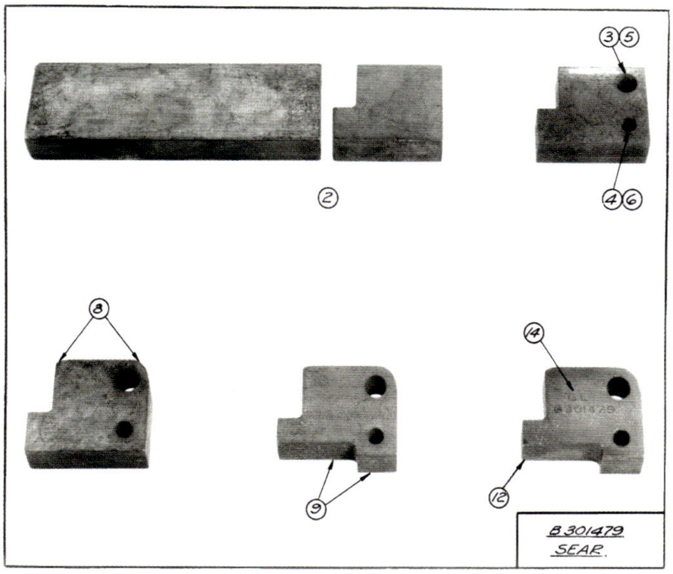

Making of the sear. This early type shown was replaced in 1954 by a reinforced design. Note the drawing number "GL B301479" stamped on the finished sear as the last work step. *Courtesy of the Rock Island Arsenal Museum*

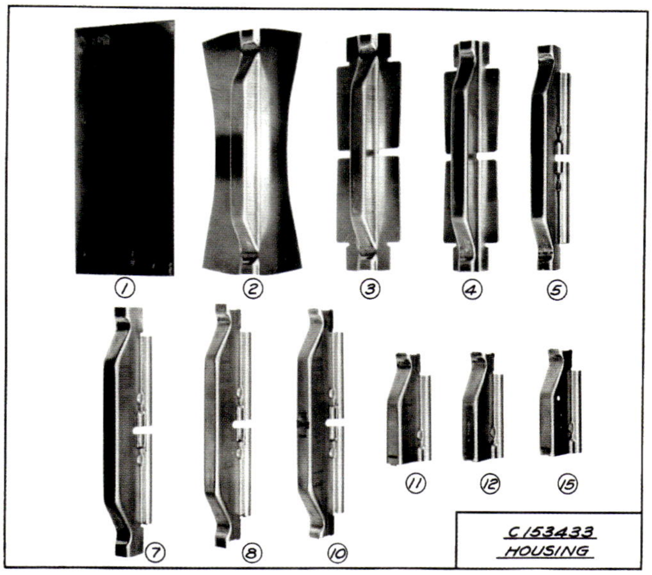

Making of the ejector housing. Two of them were pressed at once and were separated in the tenth work step. *Courtesy of the Rock Island Arsenal Museum*

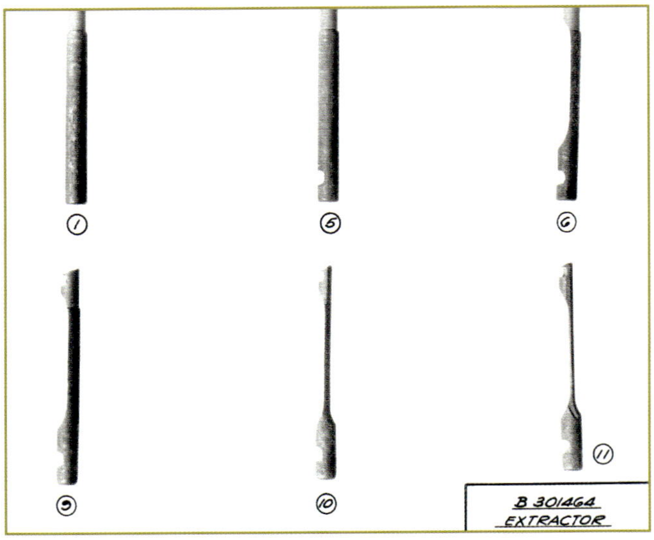

The making of the extractor was time consuming and required many work steps. *Courtesy of the Rock Island Arsenal Museum*

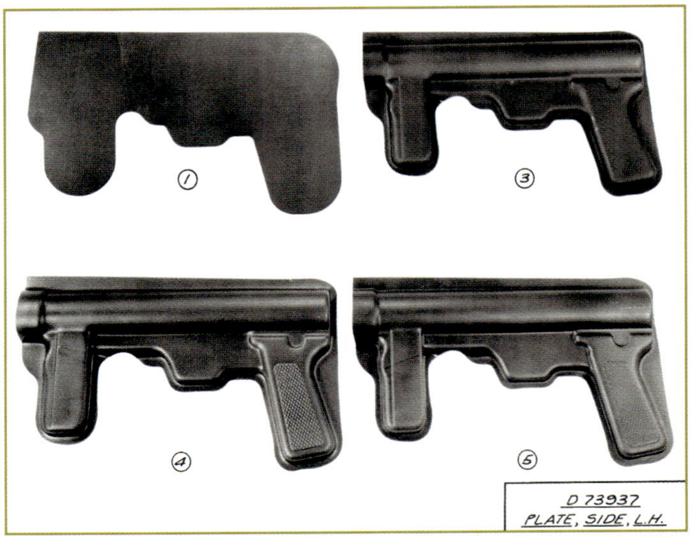

Pressing of the left receiver half. *Courtesy of the Rock Island Arsenal Museum*

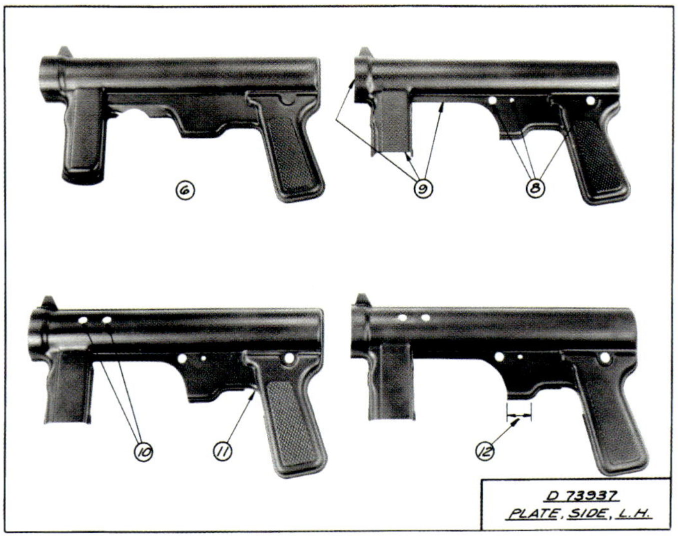

Deburring the left receiver half, drilling holes, and placing cutouts. *Courtesy of the Rock Island Arsenal Museum*

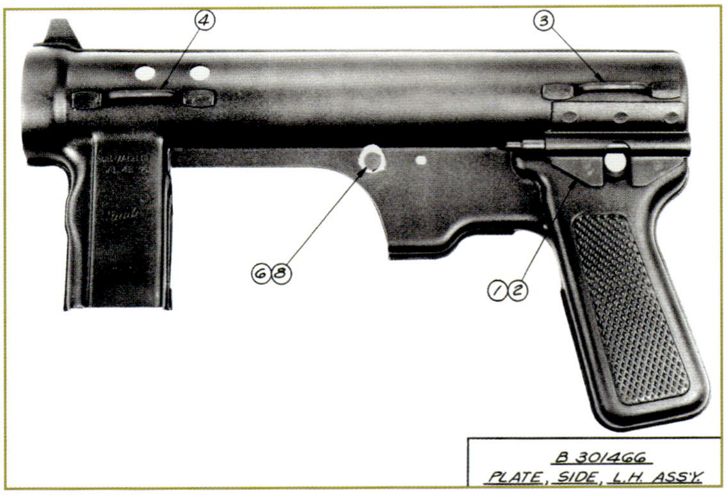

Placing the markings on the magazine well and adding sling eyes and tubular holders for the stock. Note the mandrel put into the holder as a placeholder to avoid reduction of the inside diameter of the holder during the welding process. *Courtesy of the Rock Island Arsenal Museum*

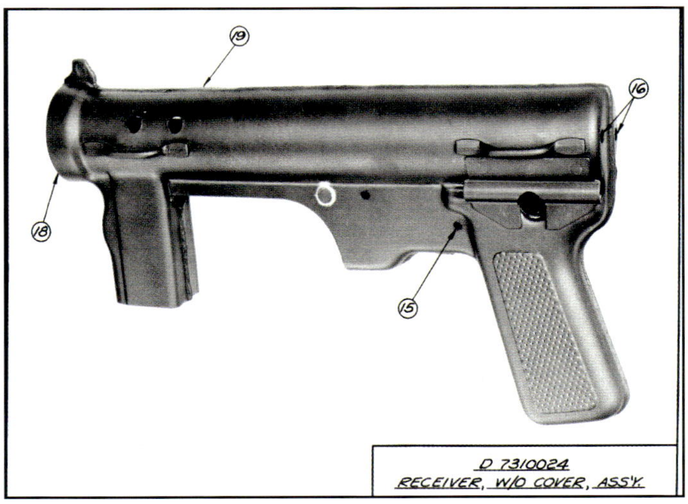

Welding both receiver halves together and drilling some more holes. *Courtesy of the Rock Island Arsenal Museum*

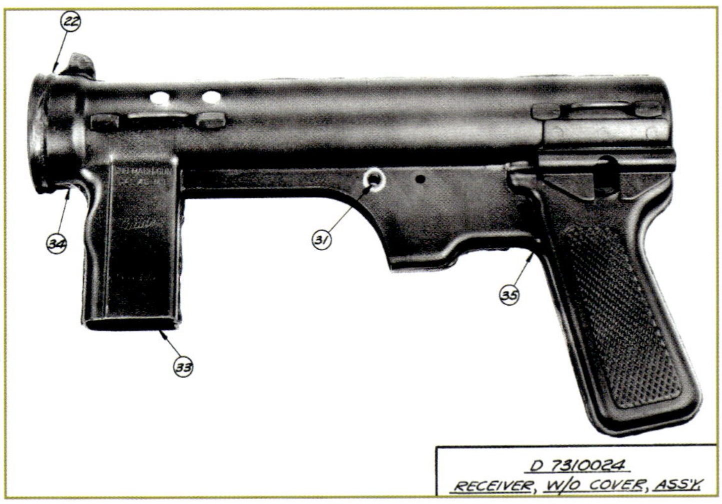

Welding the barrel bushing in place and deburring and grinding of the magazine well. *Courtesy of the Rock Island Arsenal Museum*

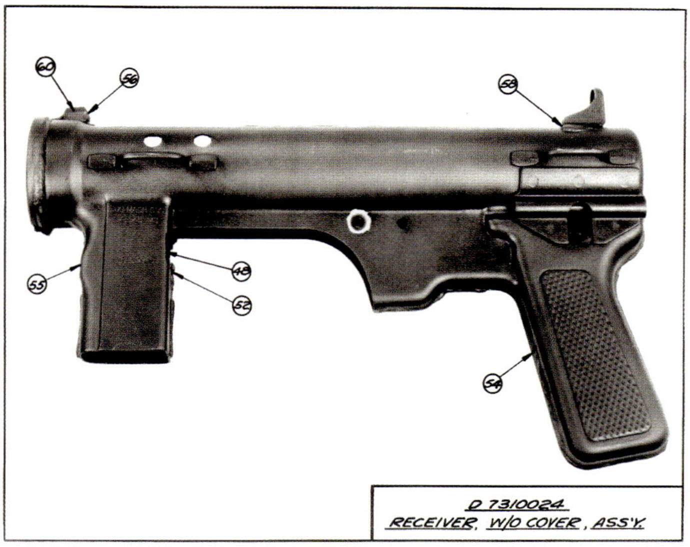

Adding the front sight and drilling the holes. *Courtesy of the Rock Island Arsenal Museum*

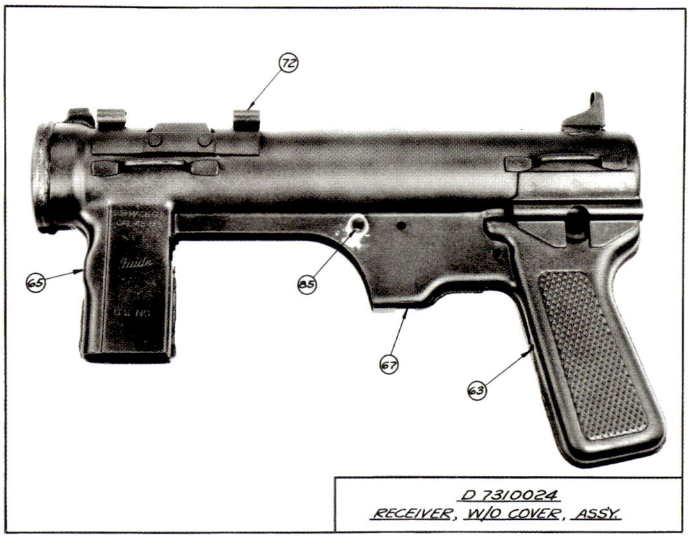

Installation of the dustcover hinge and drilling of holes. *Courtesy of the Rock Island Arsenal Museum*

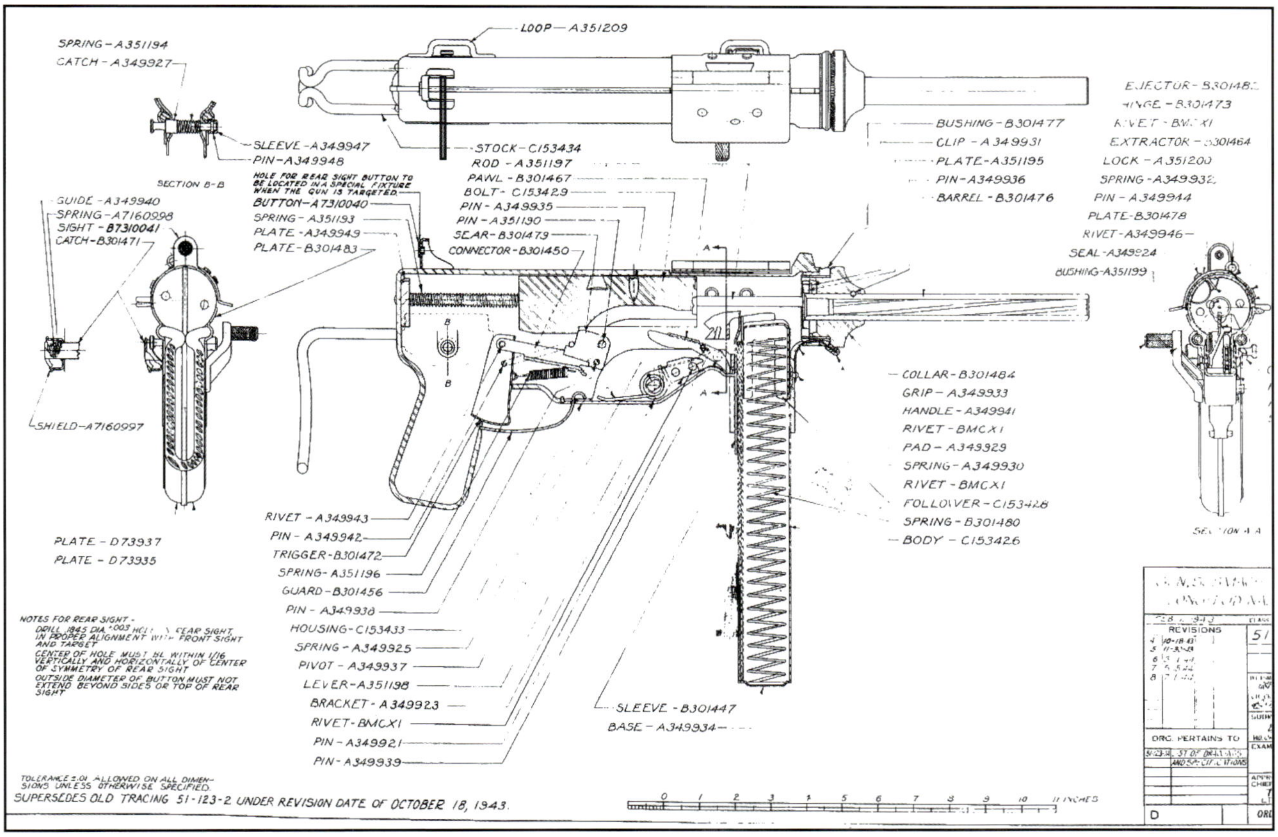

Overview of all components of the M3 and associated drawings. *Courtesy of the Rock Island Arsenal Museum*

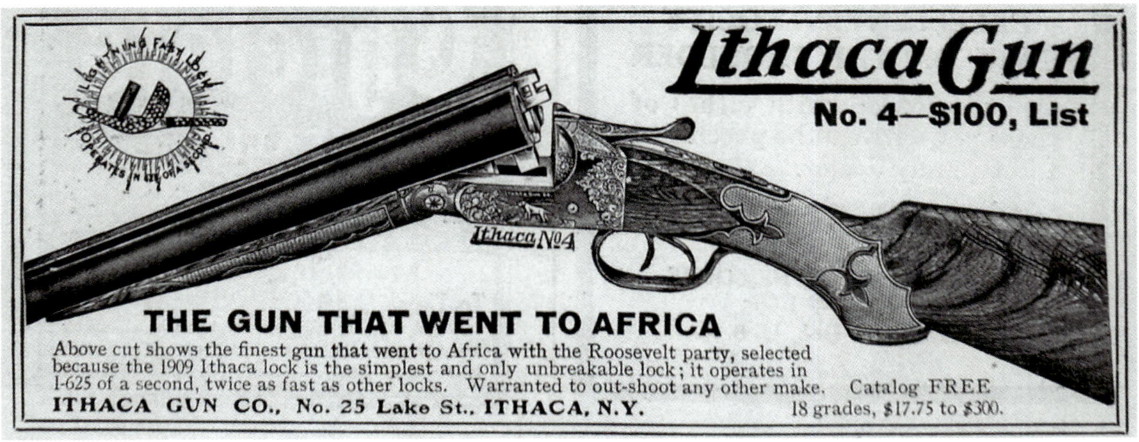

Advertisement of the Ithaca Gun Company for a hunting rifle from 1909

The Ithaca Gun Company was well known for its excellent shotguns.

An M3A1 made by Ithaca. *Courtesy of Mike Stannard*

Markings on an M3A1 made by Ithaca. *Courtesy of Mike Stannard*

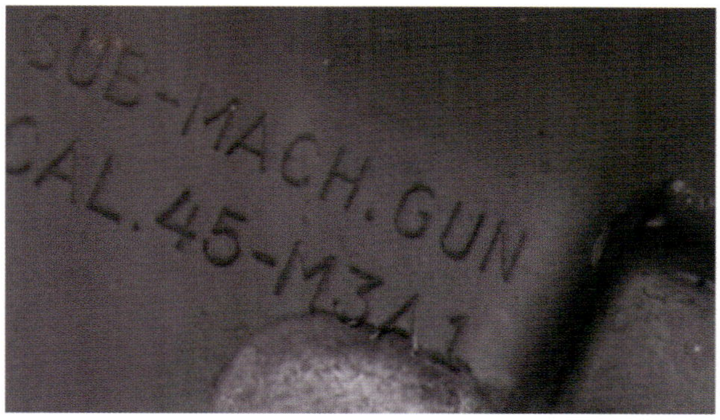

The complete marking is visible after removing the magazine release button. *Courtesy of Mike Stannard*

Comparison of the pistol grip patterns: Guide Lamp (*left*) and Ithaca (*right*)

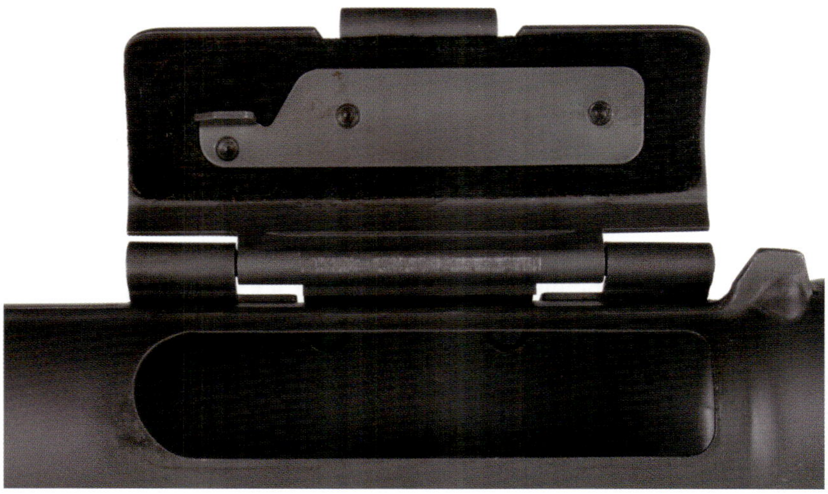

Ejection port and dustcover. *Courtesy of Mike Stannard*

Left: The oiler of an Ithaca-made M3A1

Right: Guide rod retaining plate made by Ithaca with drawing number. The abbreviation "ITG" stands for Ithaca Gun. *Courtesy of Mike Stannard*

Bottom: Traces of milling on the bolt. *Courtesy of Mike Stannard*

View over Fall Creek to the former site of the Ithaca Gun Company. The old buildings were torn down a long time ago (photo taken in autumn 2013). *Photo by Claudia Heidler*

Only the lonely chimney recalls past times. *Photo by Claudia Heidler*

Today the site is still unused because the ground is contaminated with lead. *Photo by Claudia Heidler*

08 MAINTENANCE, REPAIR, AND DESTRUCTION

EVEN A SIMPLE AND hardly susceptible weapon such as the M3/M3A1 requires a minimum of care. Thanks to its simple disassembly without tools and the easy replacement of defective parts, the cost for keeping the weapon working remains modest.

MAINTENANCE

For the weapon's use in the field, various detergents, oils, and fats were available. For thorough cleaning of stored weapons or spare parts after their delivery, a petroleum-based solvent (dry-cleaning solvent) was used. Smaller parts were bathed in it, and larger ones were wiped with it. During this work, gloves were to be used to avoid fingerprints, and of course smoking was strictly forbidden. The cleaning solvent should not come in contact with rubber parts.

After shooting the weapon, its owner should clean the barrel with "rifle bore cleaner." This liquid removes the residue in the bore and works as a rust preventive. The manuals mention that the shoulder stock, when used as a cleaning rod, was not allowed to be inserted too far into the barrel, to prevent damage of the bolt face. It was suggested to put a finger between the rods to serve as a limiter. If no bore cleaner was available, the barrel could alternatively be flushed with hot water containing dissolved soap.

To oil the gun components, three different kinds of "preservative lubricating oil" in the versions "special," "medium," and "light" were available. The "special" provided better and longer-term rust protection. The "medium" oil was for exceptional environmental influences (prior to amphibious-landing operations, extreme temperatures, or short-term storage). The "light" oil had only a limited protective effect and should be renewed daily on the moving parts. This oil was carried in the oil can of the weapon.

Two brushes were officially handed out as cleaning tools. The "Chamber Cleaning Brush M6" was used to clean the cartridge chamber and consists of a twisted metal handle with permanently attached plastic bristles. The "Caliber .45 Cleaning Brush M5," with bronze bristles for cleaning the barrel, had to be screwed on to the rods of the wire stock.

REPAIRS

The operator was allowed to disassemble his weapon only for cleaning purposes; he was not permitted to make repairs or replace any parts. Only the replacement of a damaged magazine was permitted as an "operator's repair" according to the service regulations.

In practice, however, things were often different. The soldiers carried a small supply of spare parts such as springs, pins, and extractors in pockets and bags, which were specially issued for this purpose. And since at the front an armorer was not always at hand, the soldiers often did the necessary repairs by themselves.

Information about repairs could be found in the technical manuals (TMs), of which various revised and expanded editions were issued:

- *TM 9-217 "Submachine Gun, Cal. .45, M3,"* from July 1943
- *TM 9-1217 "Ordnance Maintenance, Submachine Gun, Cal. .45, M3 and M3A1,"* from December 1945
- *TM 9-1005-229-12 "Submachine Gun, Caliber .45: M3, W/E (1005-672-1767) & Submachine Gun, Caliber .45: M3A1, W/E (1005-672-1771),"* from November 1969
- *TM 9-1005-229-35 "DS, GS, and Depot Maintenance Manual Including Repair Parts and Special Tools Lists Submachine Gun, Caliber .45, M3, W/E, Submachine Gun, Caliber .45, M3A1, W/E."* from 1969

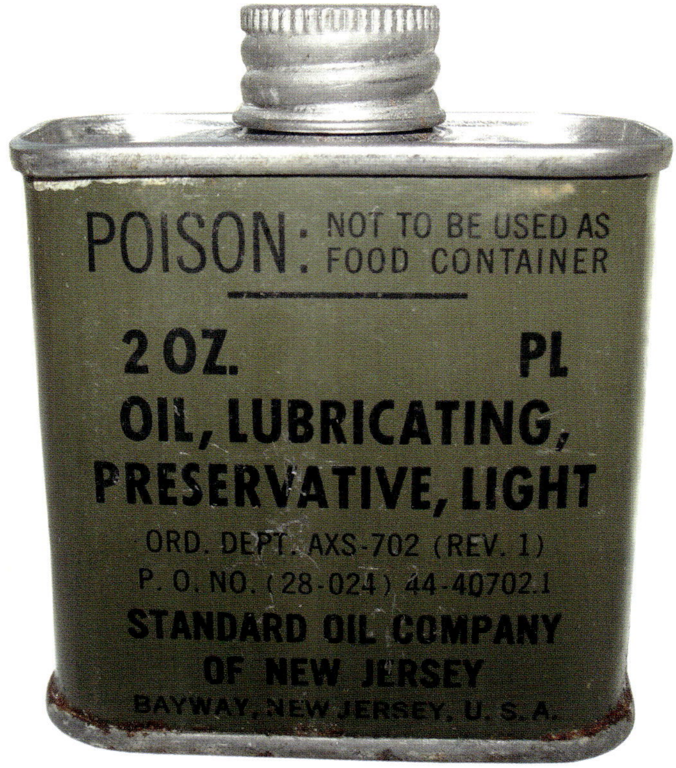

Gun oil "light," which was filled in the oiler of the weapon

Can with 2 ounces of "Cleaner Rifle Bore"

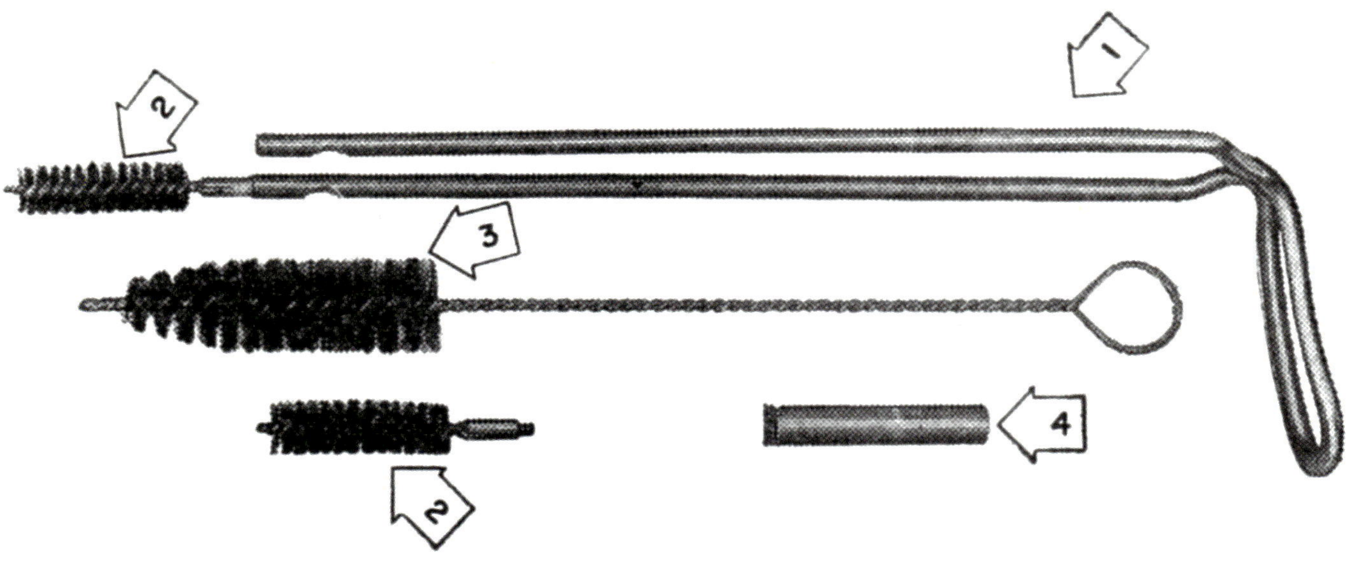

1. Stock.
2. Caliber .45 Cleaning Brush, M5.
3. Chamber Cleaning Brush, M6.
4. Oiler.

Figure 29. Cleaning tools.

Cleaning tools of the M3. Note the already improved stock ends with threads for attaching the type M2 brush.

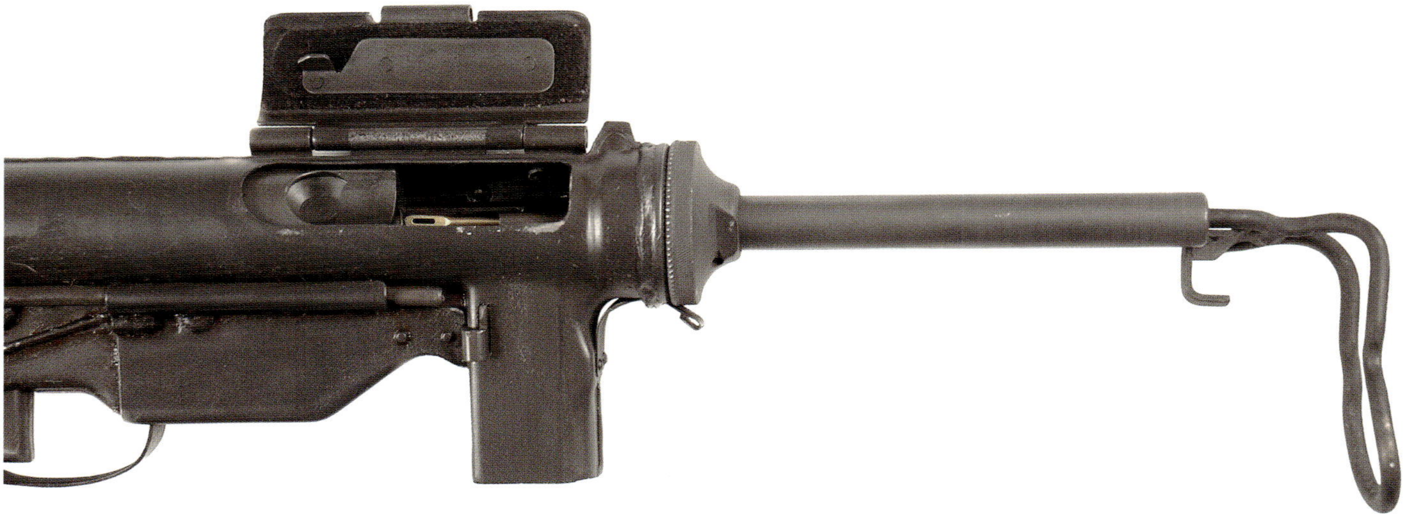

The limiter of the M3A1 stock stops the rod right before hitting the bolt face. *Courtesy of Frank Iannamico*

A broken extractor could be changed without the assistance of an armorer.

Field-repaired cocking crank of an M3. *Courtesy of Rob Stott*

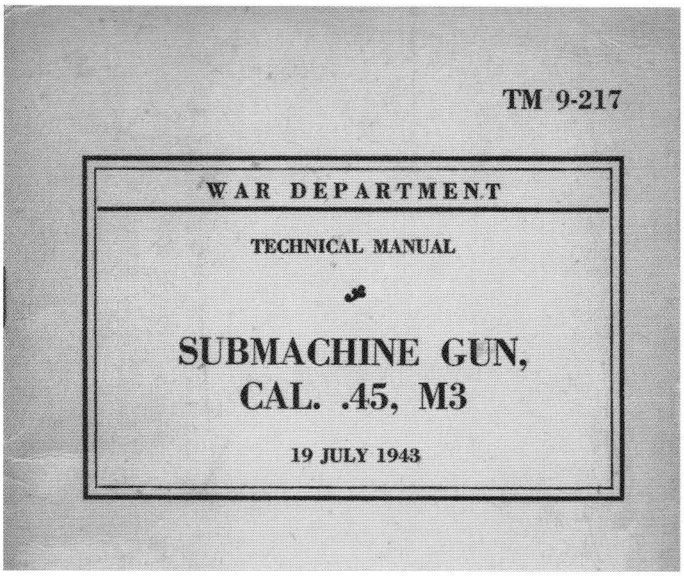

TM 9-217

WAR DEPARTMENT

TECHNICAL MANUAL

SUBMACHINE GUN,
CAL. .45, M3

19 JULY 1943

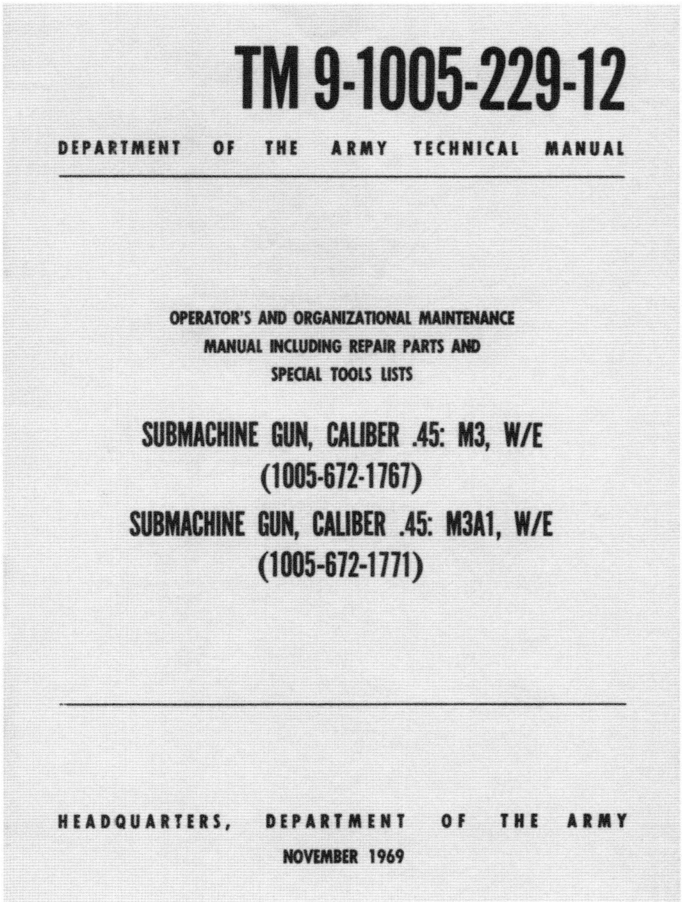

TM 9-1005-229-12

DEPARTMENT OF THE ARMY TECHNICAL MANUAL

OPERATOR'S AND ORGANIZATIONAL MAINTENANCE
MANUAL INCLUDING REPAIR PARTS AND
SPECIAL TOOLS LISTS

SUBMACHINE GUN, CALIBER .45: M3, W/E
(1005-672-1767)
SUBMACHINE GUN, CALIBER .45: M3A1, W/E
(1005-672-1771)

HEADQUARTERS, DEPARTMENT OF THE ARMY
NOVEMBER 1969

Additionally, there were technical data sheets, supplements for the technical manuals, and other information available, such as the "modification work orders" (MWOs) or the "standard nomenclature lists" (SNLs) for identification and storage of gun parts. The *SNL A-58 "Gun, Submachine, Cal. .45, M3,"* from April 26, 1943, lists each individual part that is required to assemble a complete M3. The list with the part names, stock numbers, and the required quantities was followed by picture plates showing the position of the individual parts of the weapon.

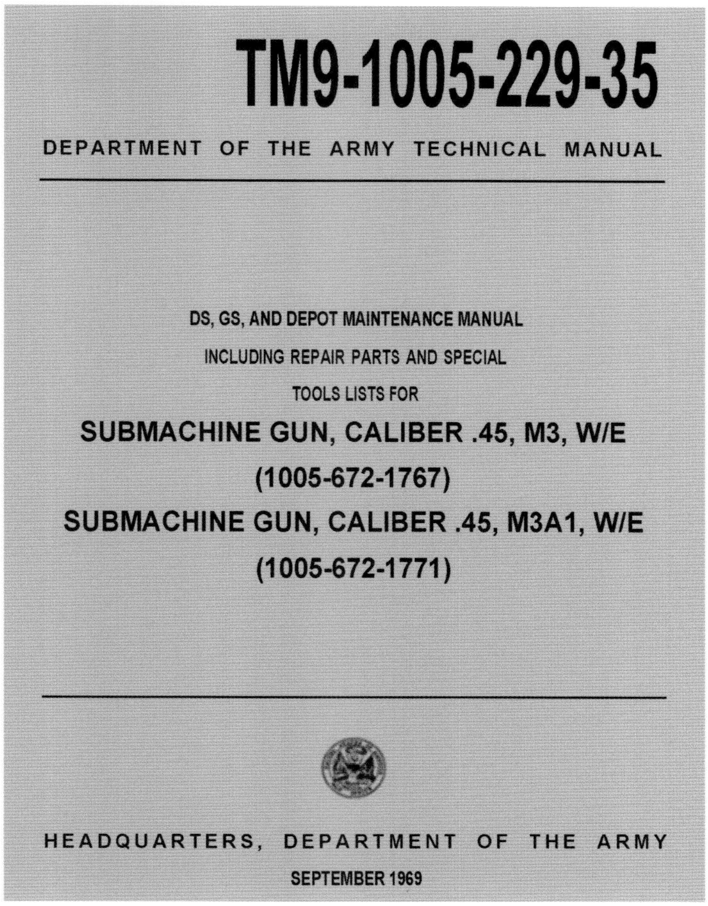

TM9-1005-229-35

DEPARTMENT OF THE ARMY TECHNICAL MANUAL

DS, GS, AND DEPOT MAINTENANCE MANUAL
INCLUDING REPAIR PARTS AND SPECIAL
TOOLS LISTS FOR
SUBMACHINE GUN, CALIBER .45, M3, W/E
(1005-672-1767)
SUBMACHINE GUN, CALIBER .45, M3A1, W/E
(1005-672-1771)

HEADQUARTERS, DEPARTMENT OF THE ARMY
SEPTEMBER 1969

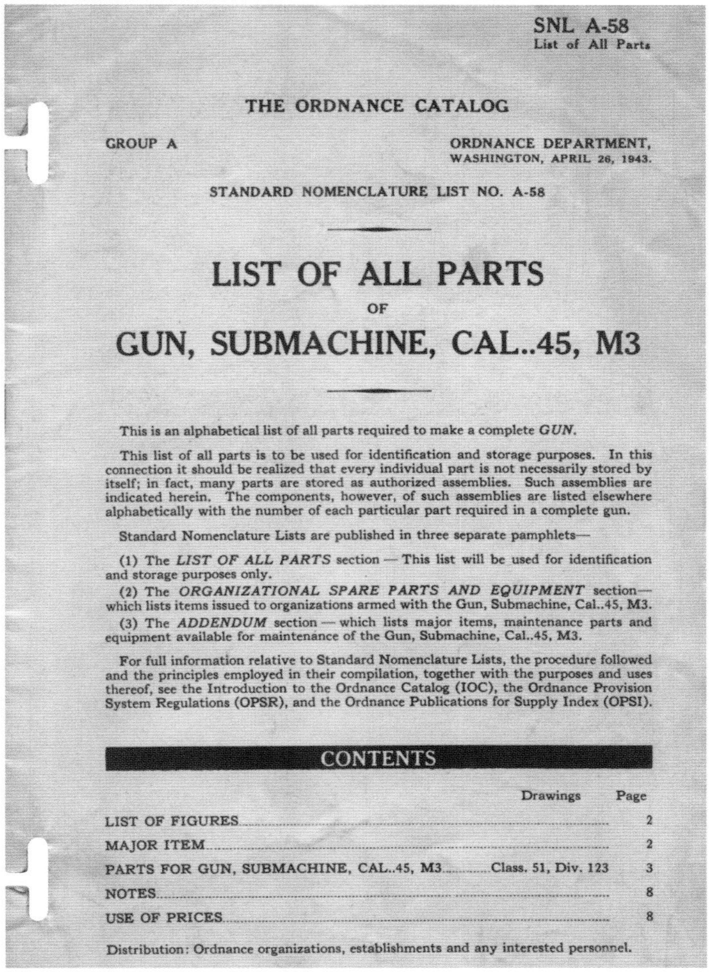

SNL A-58
List of All Parts

THE ORDNANCE CATALOG

GROUP A

ORDNANCE DEPARTMENT,
WASHINGTON, APRIL 26, 1943.

STANDARD NOMENCLATURE LIST NO. A-58

LIST OF ALL PARTS
OF
GUN, SUBMACHINE, CAL..45, M3

This is an alphabetical list of all parts required to make a complete *GUN*.

This list of all parts is to be used for identification and storage purposes. In this connection it should be realized that every individual part is not necessarily stored by itself; in fact, many parts are stored as authorized assemblies. Such assemblies are indicated herein. The components, however, of such assemblies are listed elsewhere alphabetically with the number of each particular part required in a complete gun.

Standard Nomenclature Lists are published in three separate pamphlets—

(1) The *LIST OF ALL PARTS* section — This list will be used for identification and storage purposes only.

(2) The *ORGANIZATIONAL SPARE PARTS AND EQUIPMENT* section—which lists items issued to organizations armed with the Gun, Submachine, Cal..45, M3.

(3) The *ADDENDUM* section — which lists major items, maintenance parts and equipment available for maintenance of the Gun, Submachine, Cal..45, M3.

For full information relative to Standard Nomenclature Lists, the procedure followed and the principles employed in their compilation, together with the purposes and uses thereof, see the Introduction to the Ordnance Catalog (IOC), the Ordnance Provision System Regulations (OPSR), and the Ordnance Publications for Supply Index (OPSI).

CONTENTS

Distribution: Ordnance organizations, establishments and any interested personnel.

Packages of Spare Parts

When fighting came to an end, the used weapons were usually subject to review. Damaged and worn-out weapons were returned to the depots, either for repair or to cannibalize serviceable parts. The ammunition used at that time was very corrosive, and so many weapons needed new barrels after being shot extensively. Ordnance-rebuilt weapons can be identified by additional stamps on the magazine well. Here, for example, a twice-reworked M3A1 with the stamps "RRAD" for "Red River Army Depot" and "RIA" for "Rock Island Arsenal." "EB" is the personal stamp of Ordnance Inspection Supervisor Elmer M. Bjerke, who was responsible for this work at the Rock Island Arsenal from 1947 until 1958. In 1958, all repair work from Rock Island Arsenal was handed over to the Springfield Arsenal.

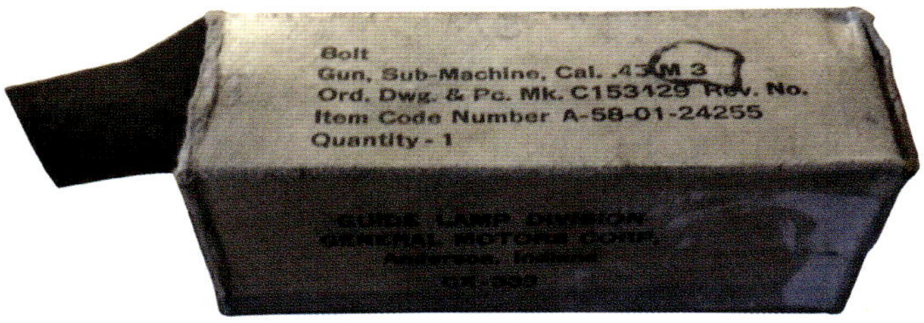

Spare bolt for the M3 with cardboard box. "Ord. Dwg. & Pc." means "Ordnance Drawing & Picture." *Courtesy of Frank C. Motley*

Springfield Armory did not manufacture bolts. This is a bolt that was hardened afterward and repacked (note the "H" stamp is for "hardened").

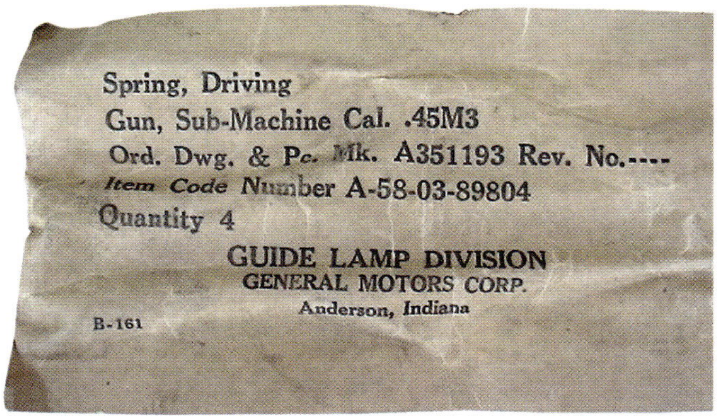

Envelope with four driving springs from wartime manufacture

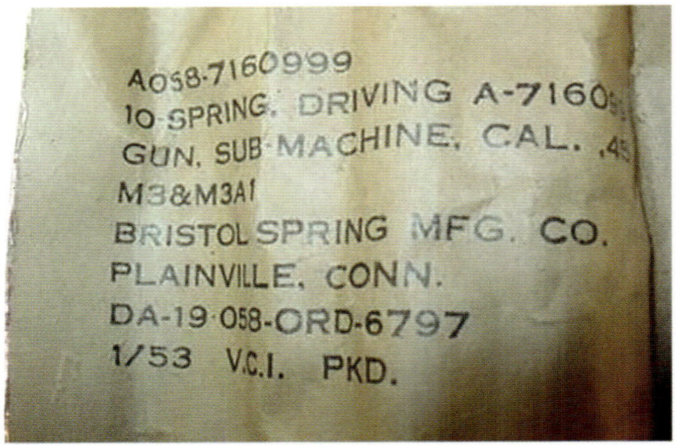

Envelope with ten driving springs. Postwar manufactured by Bristol Spring Mfg. in Plainville, Connecticut.

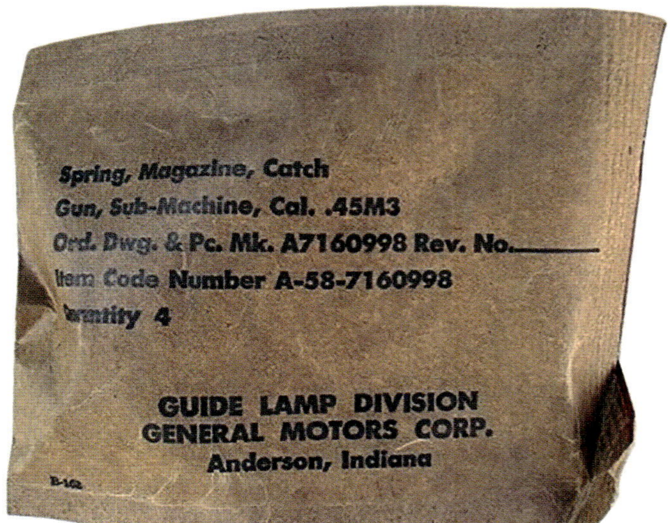

Envelope with four springs for the magazine release button

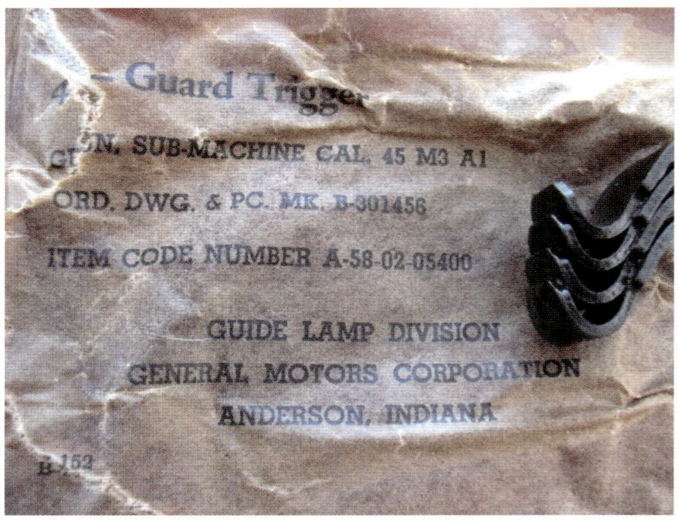

Four trigger guards with packaging made by Guide Lamp

Markings on the trigger guards with part number B 301456-4 and maker's mark GL

This M3A1 was reworked twice: "RRAD" and "RIA EB." As the sloppy stamping shows, the letters were stamped separately and not using a single punch.

Many facilities were engaged in the repair of all types of weapons. However, it is not exactly known which of them reworked the M3 or M3A1:

AA	Augusta Arsenal
AN	Anniston Arsenal
BA	Benicia Arsenal
MR	Mt. Rainier Ordnance Depot
LE	Letterkenny Ordnance Depot
OG	Ogden Arsenal
RA	Raritan Arsenal
RIA FK	Rock Island Arsenal / Frank Krack (1941–1946)
RIA EB	Rock Island Arsenal / Elmer Milton Bjerke (1947–1958)
RRA	Red River Arsenal (in the '60s, renamed in Army Depot)
RRAD	Red River Army Depot
SA	Springfield Arsenal
SAA	San Antonia Arsenal
TE	Tooele Ordnance Depot

Another M3 that was reworked at the Rock Island Arsenal. *Courtesy of Ian Philips*

Elmer Bjerke (*on right at the rear*) inspecting Browning machine guns at the Rock Island Shop K on January 7, 1943. *Courtesy of the Rock Island Arsenal Museum*

Two M3s with "O.G." stamping of the Ogden Arsenals in Utah. As the doubling shows, the letters were stamped together with one single punch. *Courtesy of Frank Goepfert / www.atfmachinegun.com*

DESTRUCTION

In order to avoid operational weapons falling into the hands of the enemy, the manuals include instructions for destroying the M3/M3A1. Basically, the destruction should be carried out as thoroughly as possible. In an emergency, the destruction had to be limited to the most-important parts, starting with those that would be the most difficult to duplicate by the enemy. The same essential parts of each weapon had to be destroyed to prevent the assembly of a functional weapon from several damaged ones.

Before reaching the combat zone, the soldiers had to be trained to quickly and adequately destroy their individual weapons in an established and uniform sequence. The following note from the 1949-dated manual is quite amusing: "Training will not involve the actual destruction of materiel."

The parts of the M3/M3A1 were removed in the following order and made unusable: bolt with guide rods and springs, cocking mechanism (only on M3s), extractor, barrel, receiver. The receiver was to be bent by smashing it against a hard object. "Bend barrel by the same method" is written in the manual from 1949, but this, however, was, without doubt, hard to do because the barrel was extremely stable. In later manuals, it is therefore proposed only to make the thread of the barrel nut unusable.

The ammunition should be destroyed as well. The best way was burning it in a large stack with enough fuel to reach the highest possible temperature. After igniting it, all people were to take cover. In most cases, thirty to sixty minutes was required to destroy the ammunition carried by small combat units.

09 TRAINING ON THE WEAPON

FOR EDUCATIONAL PURPOSES, VARIOUS "field manuals" (FM) for the M3 and M3A1 were available. They were revised and reprinted several times during the period of use of the weapon. For older issues, supplementary sheets were distributed to glue into the manuals:

Shooting of various weapons during training (M3, M1 carbine, and Colt 1911). *National Archives*

Counting the hits. *National Archives*

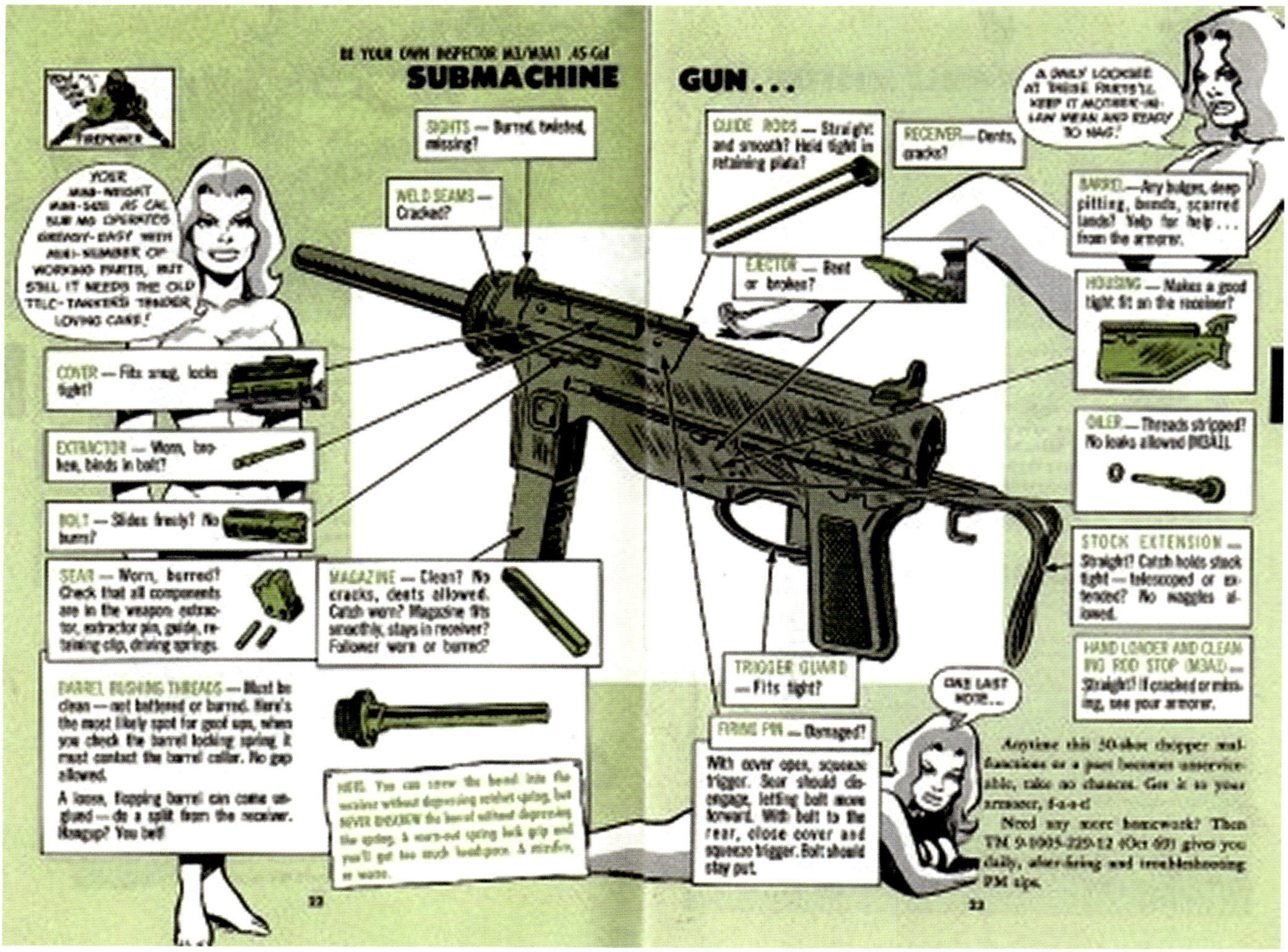

Short instruction in comic style

- *FM 23-41 "Submachine Gun Caliber .45, M3,"* from October 1943
- *FM 23-41 "Submachine Gun Caliber .45, M3 and M3A1,"* from August 1949
- *FM 23-41 "Submachine Guns Caliber .45 M3 and M3A1,"* from July 1957
- *FM 23-41 "Submachine Guns, Caliber .45 M3 and M3A1,"* from June 1974

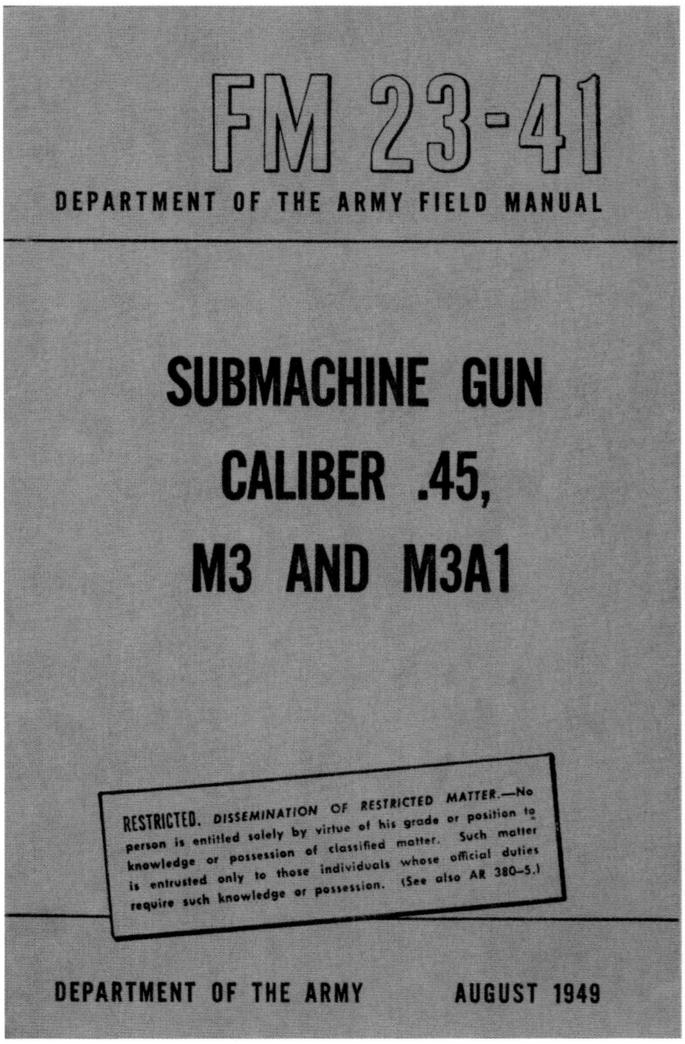

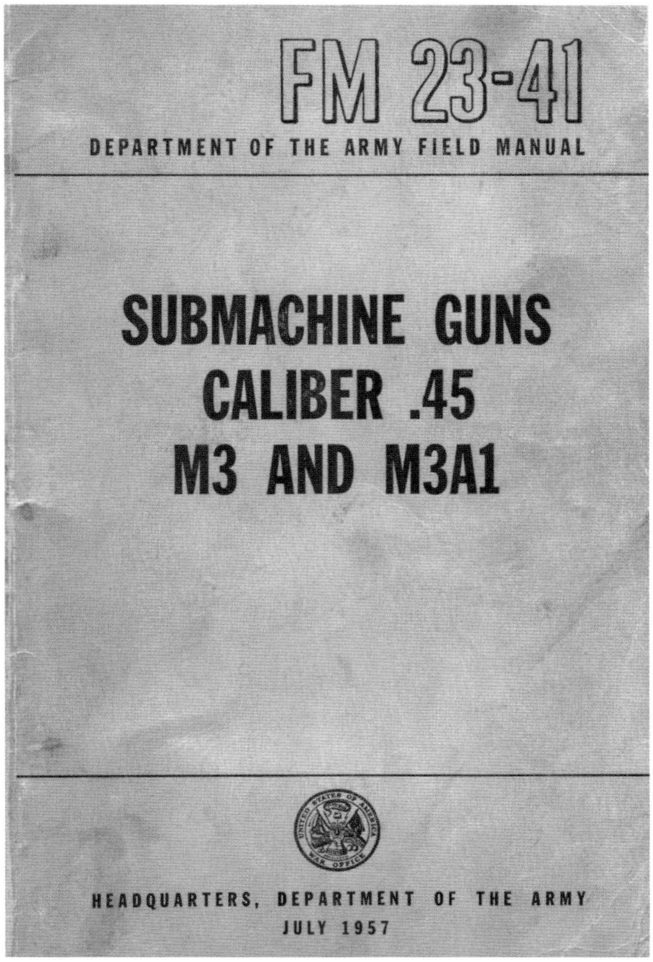

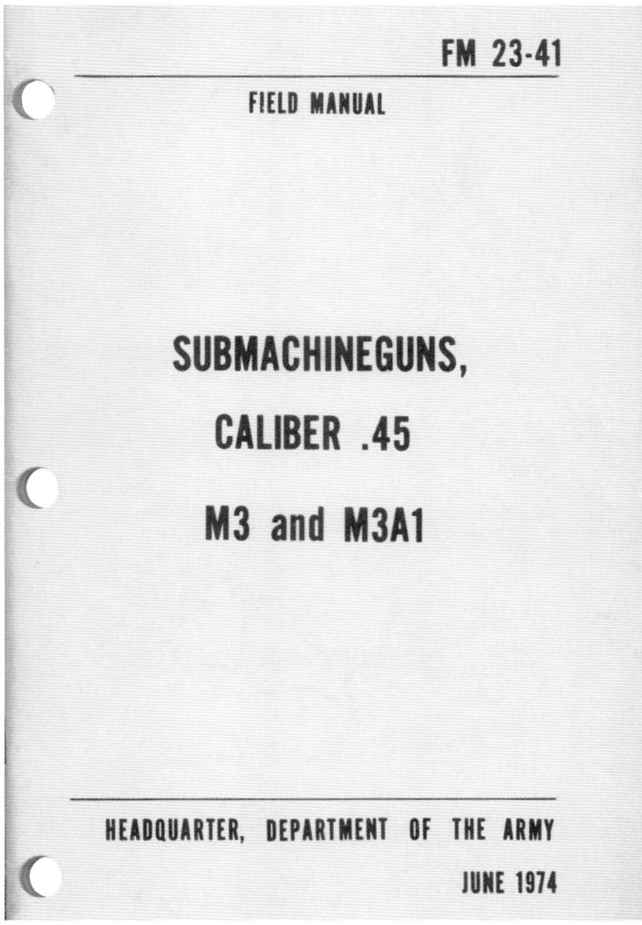

The new M14 rifle should replace all the weapons on the table: the BAR M19A2, M1 Garand, M2 carbine, and M3A1. *National Archives*

CARRYING POSITION | RAISE ARMS | PORT ARMS | INSPECTION ARMS | PRESENT ARMS

Figure 30. Manual of arms.

From FM 23-41, dated 1949

Proper handling of the weapon during drill is accurately described in the manuals. Some executions of the commands have thereby changed somewhat over the years. Same with the shooting positions: some have slightly changed, as can been seen in the figures.

From FM 23-41, dated 1974

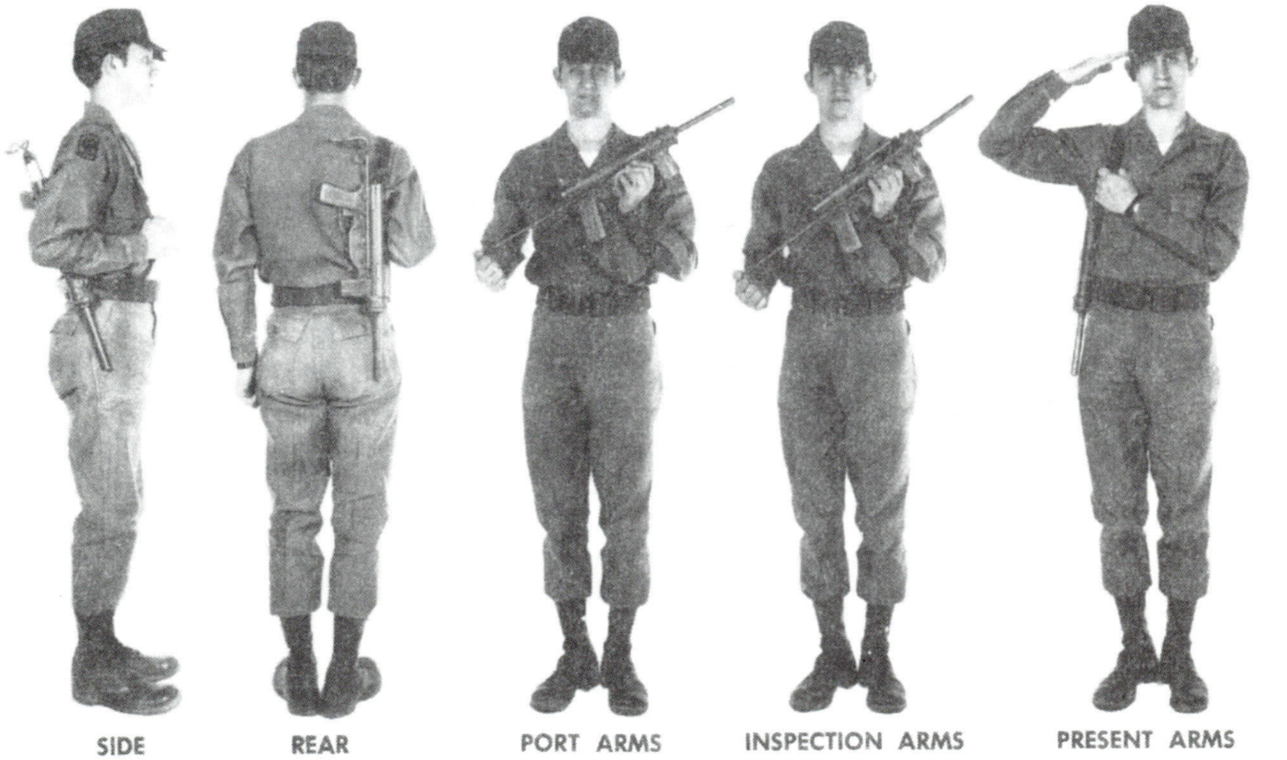

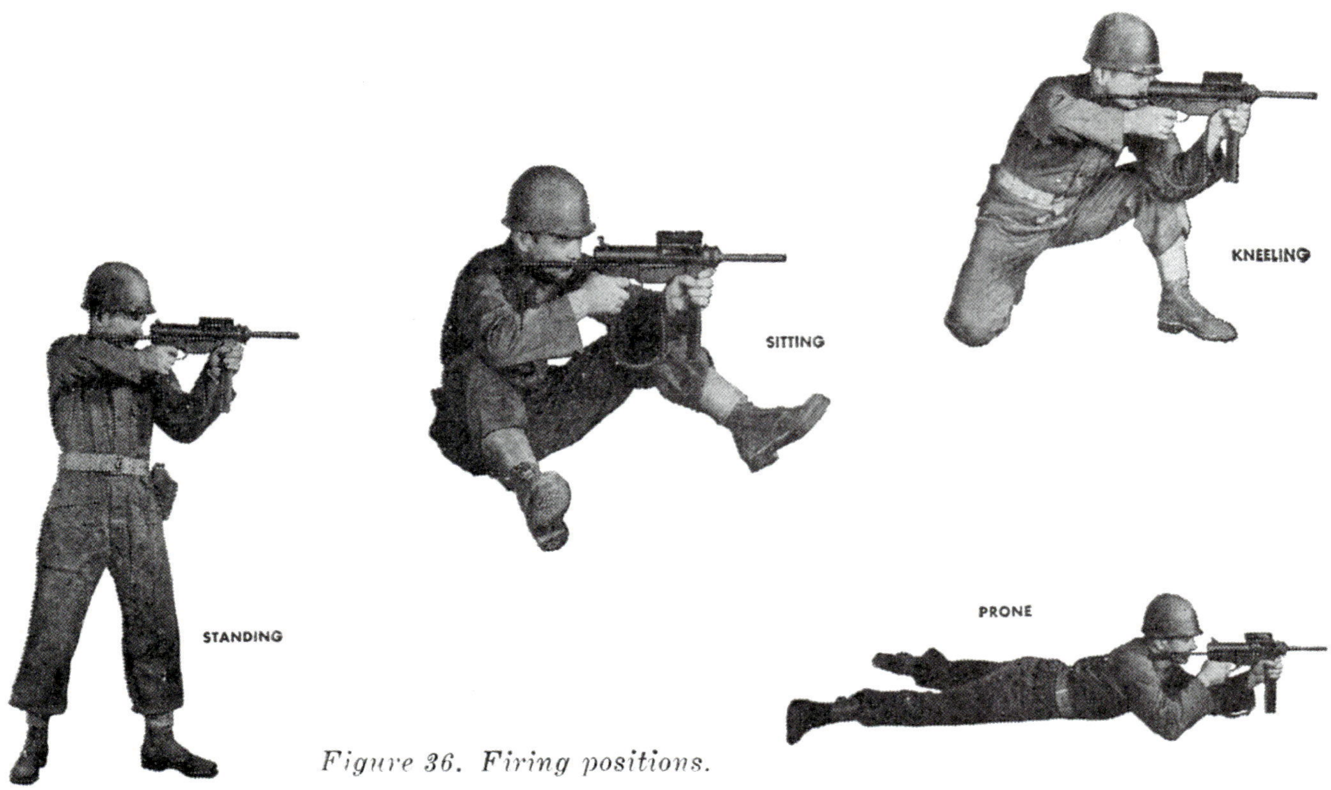

Figure 36. Firing positions.

From FM 23-41, dated 1949

Figure 30①. Firing positions.

From FM 23-41, dated 1974

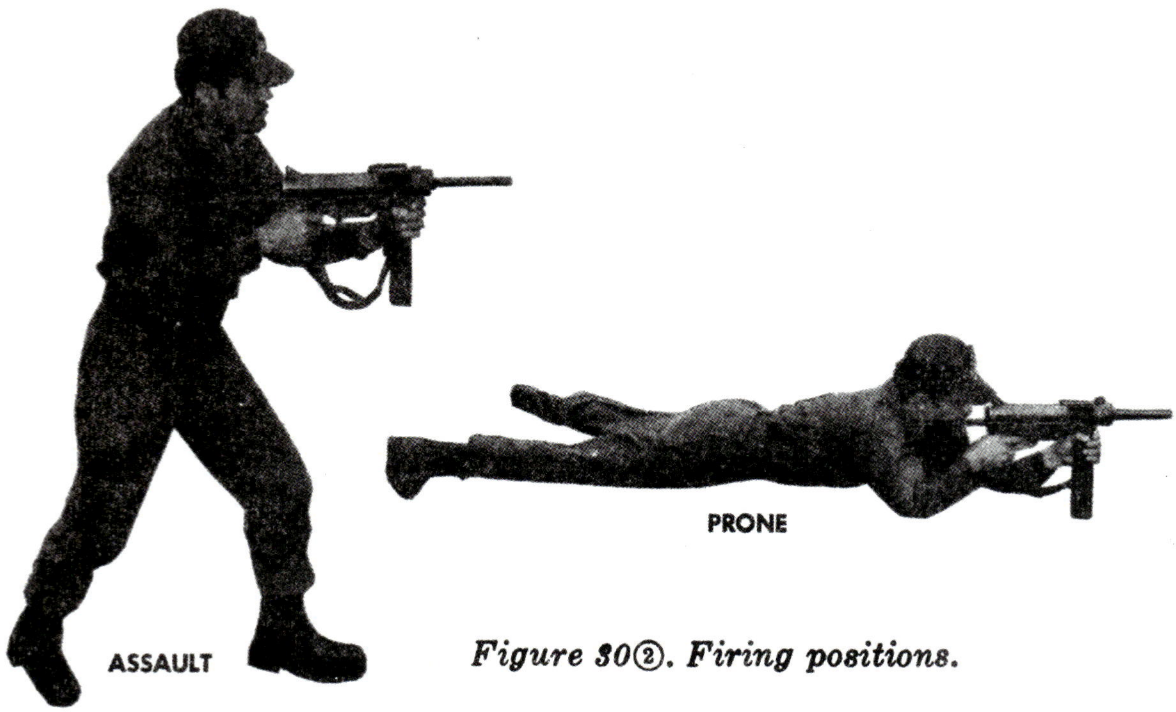

Figure 30②. Firing positions.

From FM 23-41, dated 1974

After all the theory, the practical training followed. In addition to the detailed explanations of operation, individual parts, and troubleshooting, the field manuals contain various exercises for the shooting training. One of the first steps was the "trigger manipulation," in which the release of single shots was practiced. Thanks to the low cycle rate of the M3 and M3A1, that skill was easy to learn.

Naturally, great emphasis was placed on a proper aiming and good shooting results. For that purpose, the standard target types E, F, and M could be used in various ways as needed: stationary, pivot type, hinge type, and sliding type. Depending on the equipment of the training center, the movable targets were more or less elaborately manufactured. And since the submachine gun should be primarily used by vehicle crews, shooting from moving vehicles was also practiced.

Extract from the training schedule

P-Period
H-Hours

Training Schedule

P	H	Subject	Text reference (this manual)	Area	Equipment and training aids
1	1	Characteristics, general data, nomenclature, dissassembly and assembly of groups.	Pars. 3–13	Classroom	Each man: Submachine gun, screw driver, waste, table, chair. Instructor: Do, plus Graphic Training Aid 9–3.
2	1	Detailed disassembly and assembly.	Pars. 8–13	do	Do.
3	2	Functioning	Pars. 24–27	do	Do.
4	1	Malfunctions, stoppages and immediate action, care and cleaning.	Pars. 14–23, 28, 29.	do	Do, plus dummy ammunition, cleaning materials.
5	1	Examination	All preceding	Classroom	Pencil, paper.
6	1	Orientation on marksmanship, first sighting and aiming exercises	Pars. 63–64, 78–83, app.	Training area.	Sighting bar.
7	2	Second sighting and aiming exercise at 50 feet.	do	do	Submachine gun, submachine gun rest, sighting disc, blackening material for sights.
8	1	Third sighting and aiming exercise at 50 feet.	do	do	Do, plus pencil and paper.
9	1	Review, continuation of third sighting exercise.	Pars. 63–64, 78–83.	do	Do.
10	1	Trigger squeeze exercise, calling the shot, breathing.	Par. 66	do	Submachine gun.
11	1	Range estimation exercises	Pars. 67, 97, 6, app.	Range	None.
12	1	Firing positions—standing sitting.	Par. 65	Training area.	Submachine gun.
13	1	Firing positions—kneeling, prone.	do	do	Do.
14	2	Marksmanship exercises	Par. 68	Range	All range equipment.
15	2	Examination	All preceding, plus par. 69.	Training area.	Pencil, paper.

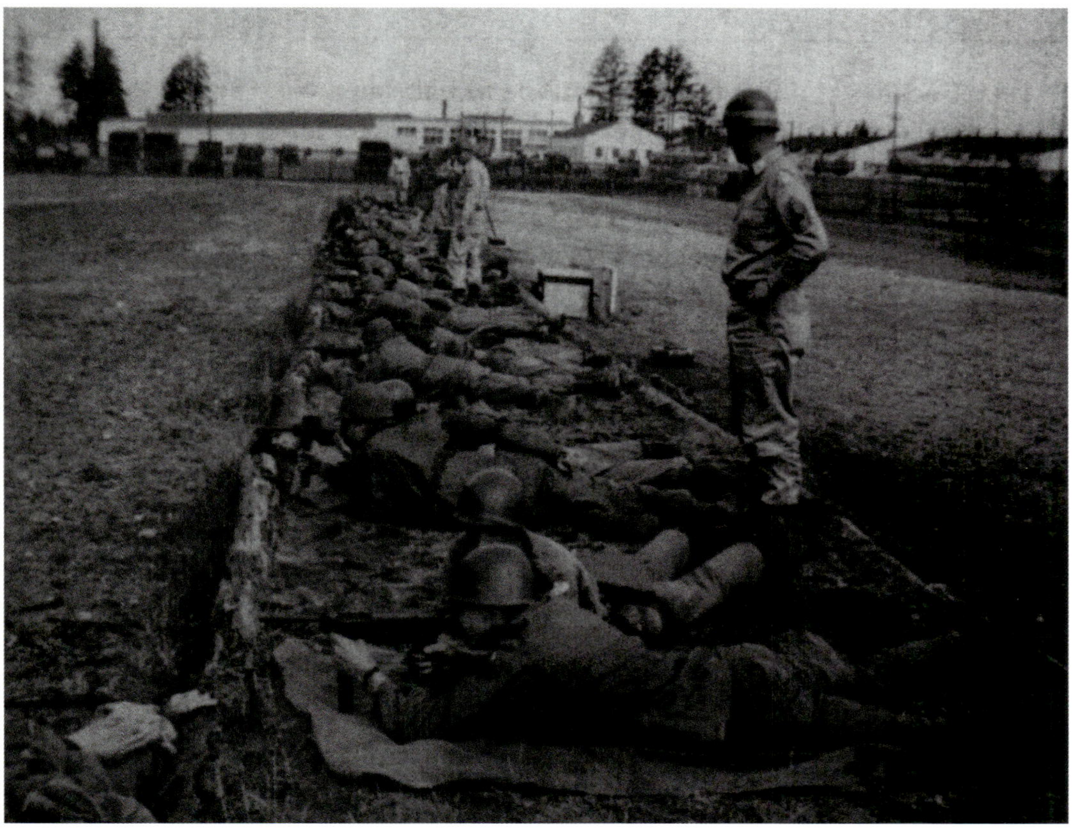

When one was shooting in the prone position, the magazine could be used as a support.

Training of an attack with machine guns and M3s.

Members of the "CASA" (Civil Affairs Holding & Staging Area) are training with the M3 somewhere in Asia in 1945. *National Archives*

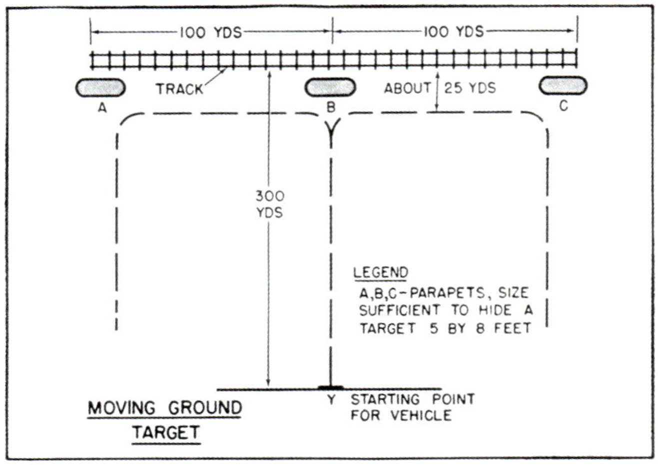

Figure 40. Moving ground target.

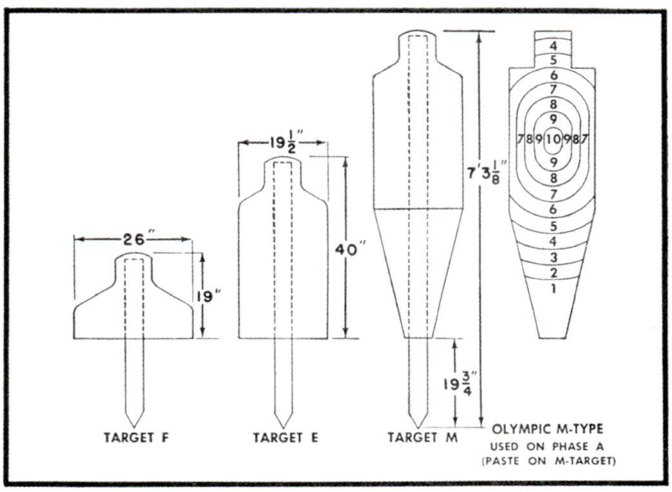

Figure 38. Targets.

The standard targets, F, E, and M

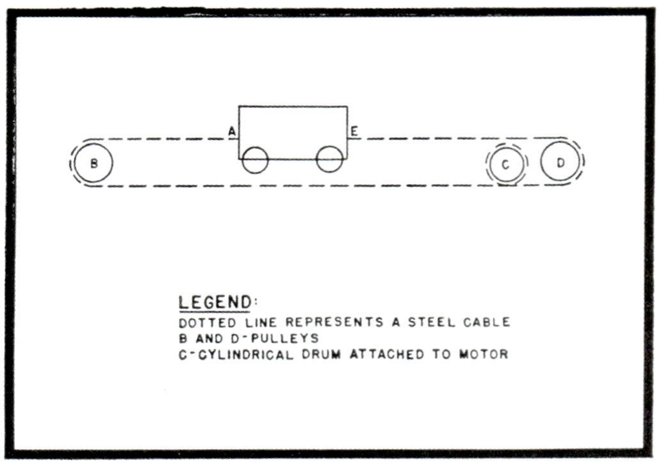

Figure 41. Moving target set-up.

Instructions for setting up an exercise for shooting on moving targets

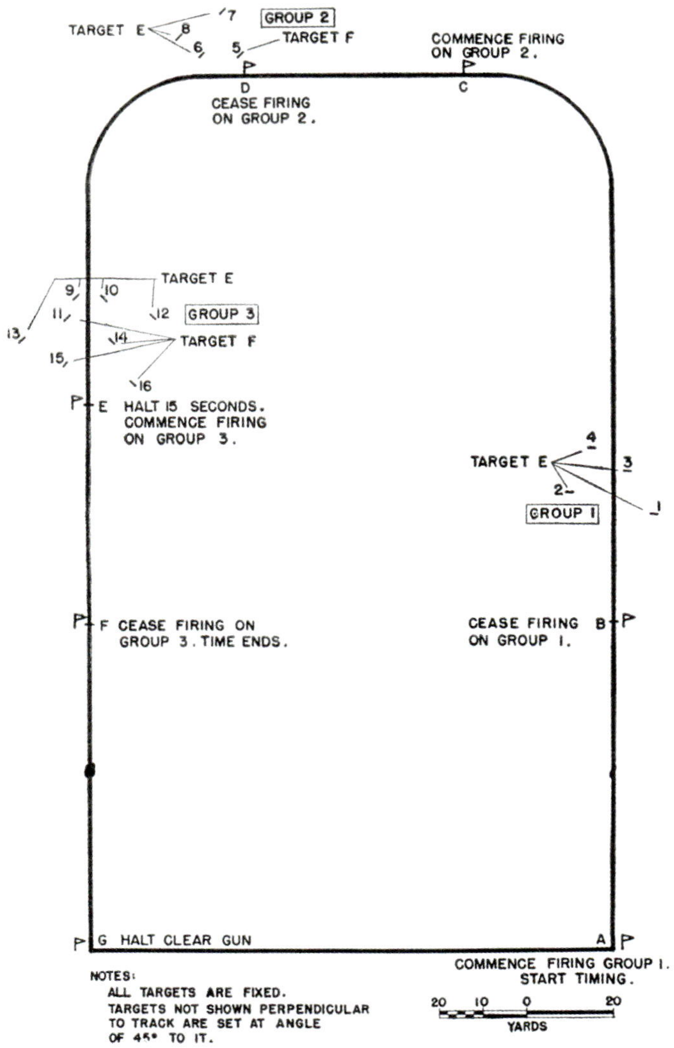

Course for firing from open vehicles

The training was supported by various training aids, which were partly to be made by the training units themselves:

DISPLAY BOARDS

There were no officially issued display boards, but the units used pictures from the manuals in enlarged form.

DEVICES FOR AIMING EXERCISES

Although the submachine gun has an effective range of 200 yards, according to the manual, only the best shooters were able to hit a target precisely at a distance of 100 yards. The sights of the M3 and M3A1 are not adjustable. For learning proper aiming without wasting ammunition, the training units could make a "sighting bar" from wood: a peephole rear sight is centrally mounted, a foresight at the "muzzle," and an eyepiece at the rear end. The eyepiece ensures that each user has the same distance from the eye to the peephole rear sight and that the results are therefore comparable. The shooter had to move the device in the correct position to aim at the target, and the instructor then checked the result.

Upon successful completion of this exercise, something similar followed with the real weapon. For this, the submachine gun was placed on a support (e.g., a wooden box) and then precisely aligned by the instructor on a movable target. Subsequently, the target was shifted slightly by a second person. Now the shooter had to aim with the rear sight (without touching the gun) and give instructions regarding in which direction the target should be moved so that it again was exactly in the center of the sight. The instructor then took the position of the shooter and checked the result. Finally, he explained to the shooter how well he did or which kind of mistakes he had made.

Gen. Donald McGowan inspects the theory lessons at Paterson Armory, New Jersey. *Courtesy of the National Guard Militia Museum of New Jersey*

Instruction for making a "sighting bar"

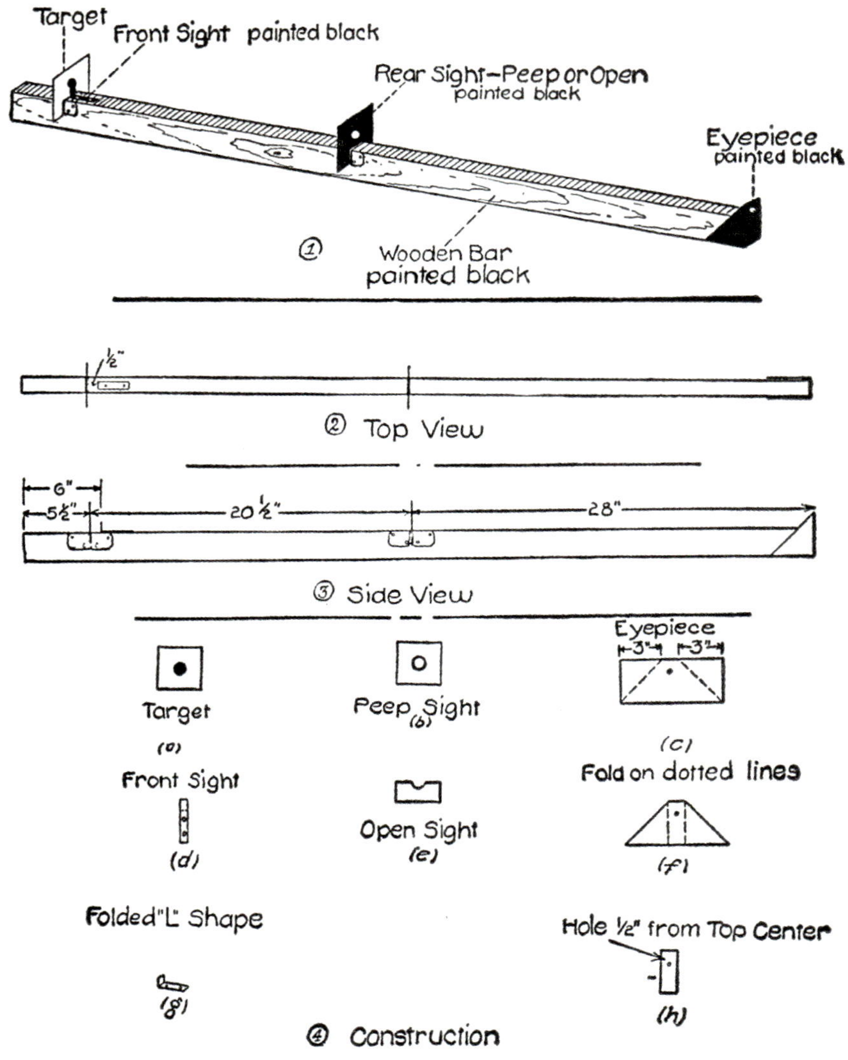

Figure 33. Sighting bar.

How to discover aiming mistakes

INCORRECT

A

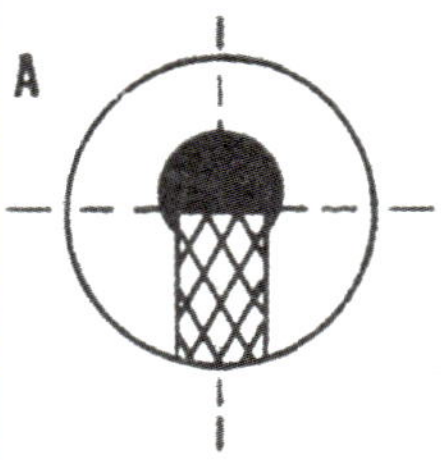

Front and rear sights properly aligned but held too high into the bull's-eye.

Result: Shot goes high.

B

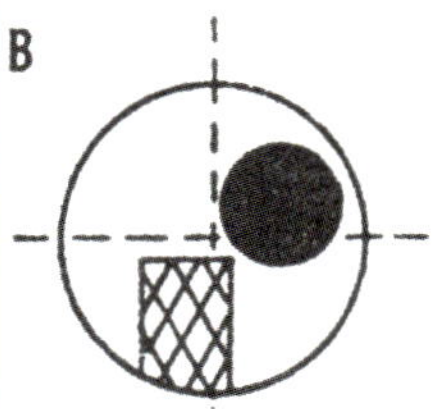

Bull's-eye below the horizontal diameter of peep sight, and front sight held left of bull's-eye.

Result: Shot goes left.

C

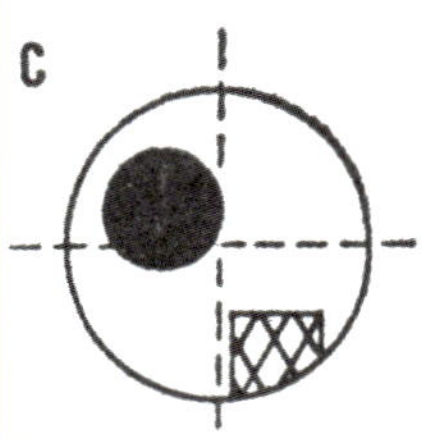

Bull's-eye below the horizontal diameter of peep sight, front sight held low and right of bull's-eye.

Result: Shot goes low and right.

CORRECT

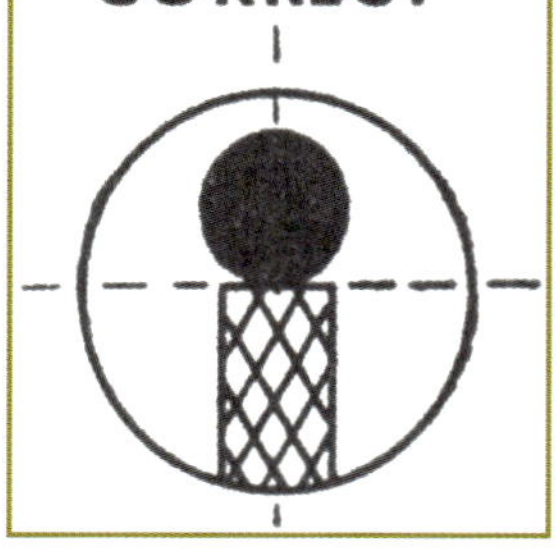

Figure 35. Submachine-gun rest and sighting disk.

Another way for training the correct aiming

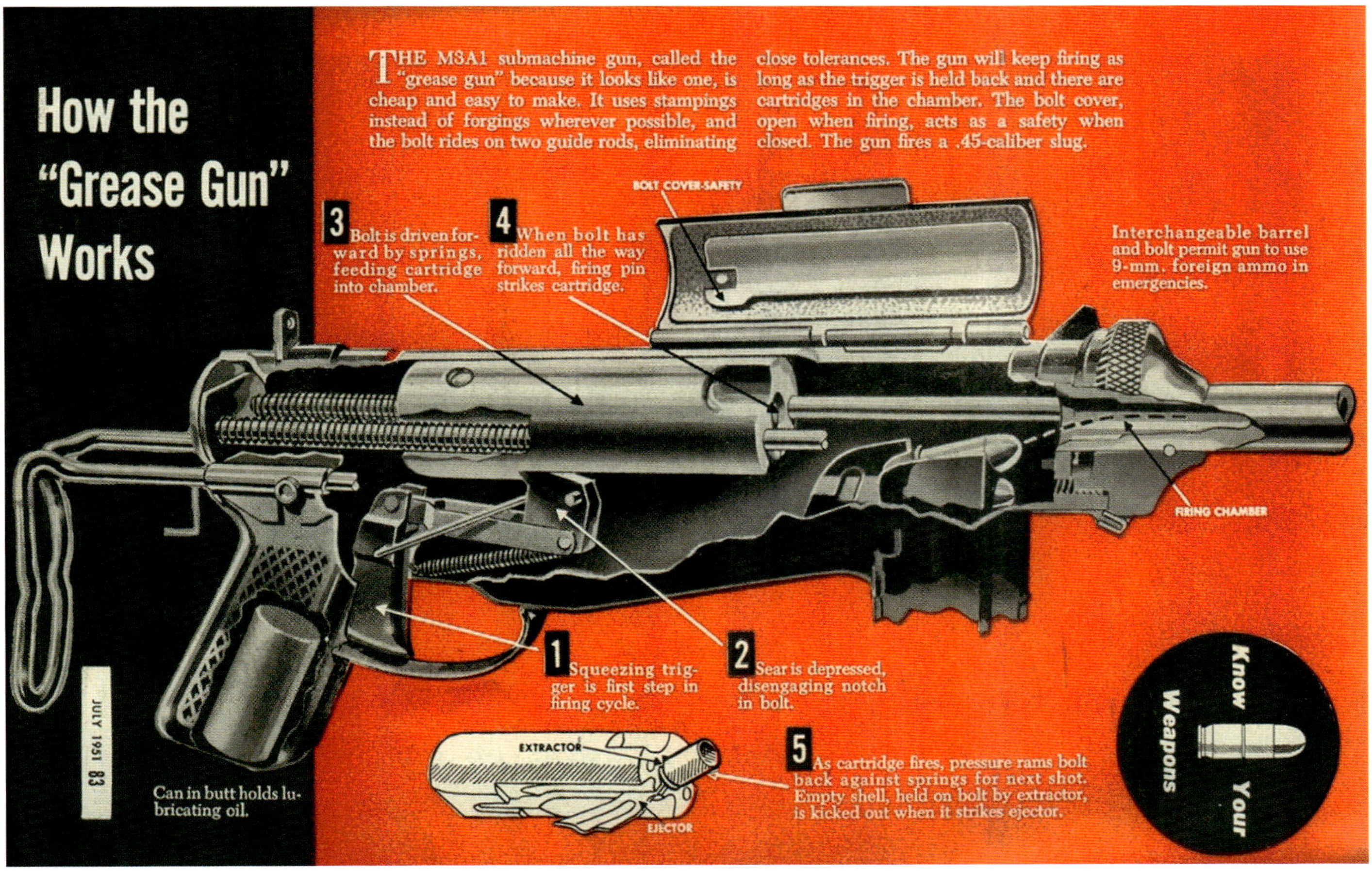

SECTIONAL MODELS

Some sectional models of the M3 and M3A1 were probably used for training purposes. However, it is unclear who had prepared these weapons and when it was done. No official instruction is known. Presumably, unserviceable weapons were used for these purposes and were cut by the training units themselves.

A sectional model of the M3 shows how the internal parts are moving. *Courtesy of the Springfield Armory NHS / US NPS*

WOODEN MODELS

According to the manuals, sectional models with moving parts should be made of wood to explain to the soldiers vividly what is going on inside the gun. Alternatively, the use of cardboard or even paper was possible, but these models were sensitive and did not last long. The models showed all the necessary internal parts, and the feeding and ejection of the cartridge could be demonstrated. A contrasting paint and markings on the parts were recommended. Mostly, these models were not made in the original dimensions, but larger. Depending on the technical skills, the models were more or less elaborately manufactured.

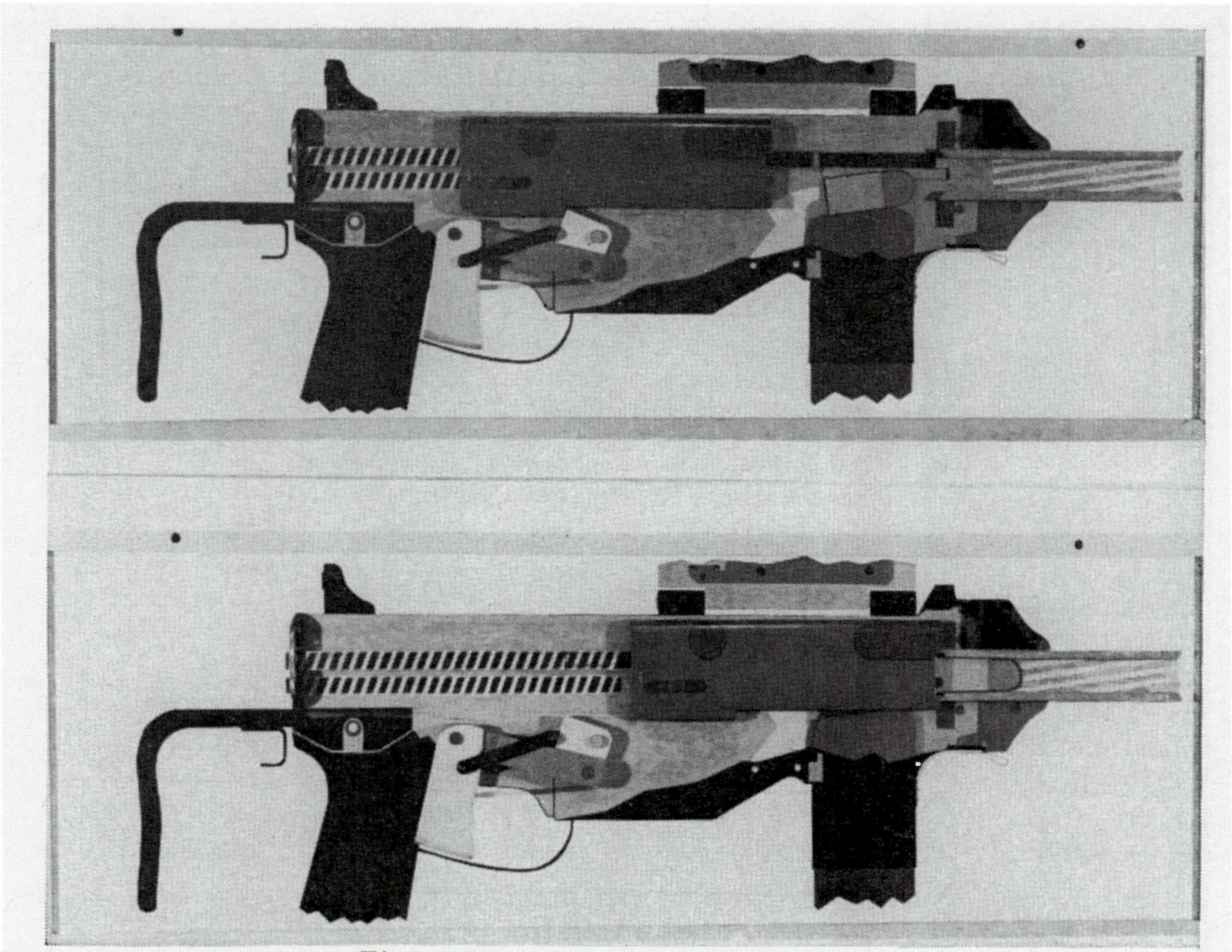
Figure 48. A wooden working model.

Wooden model in enlarged size, with moving parts pictured in the FM 23-41, from 1957

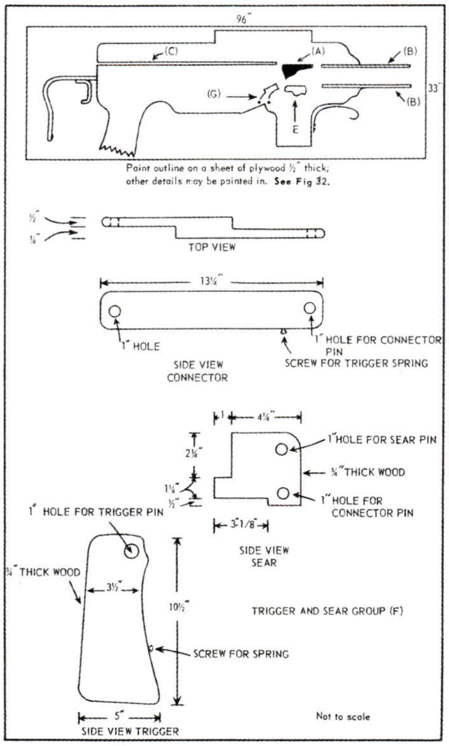

Figure 33①. Diagram of working model.

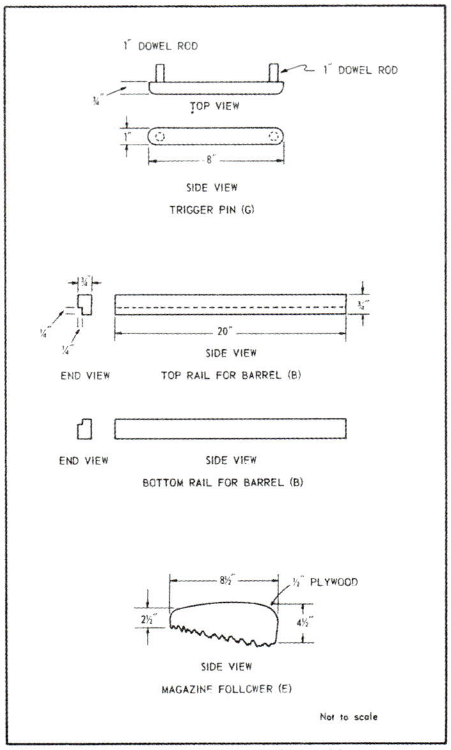

Figure 33②. Diagram of working model.

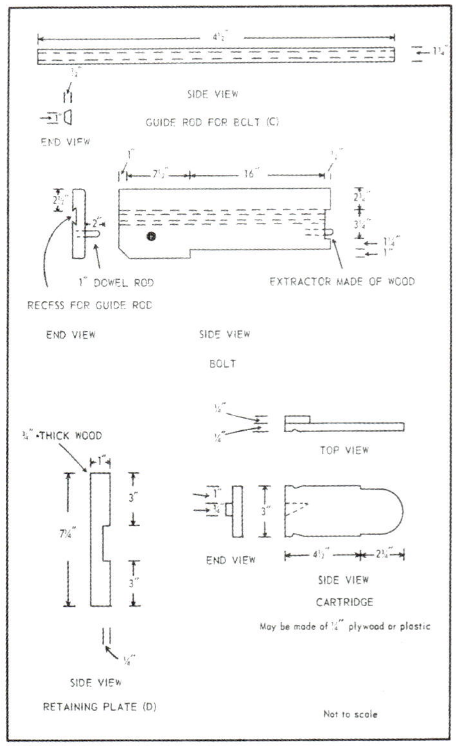

Figure 33③. Diagram of working model.

Instruction for making a wooden model with moving parts in FM 23-41, from 1974

RUBBER MODELS

During the Vietnam War, some training units used reproductions of the M3A1 made from rubber. A mold was taken of a weapon and then poured out. Depending on the care given, these casts looked very real, and even the markings were often visible. For stability reasons, the barrels were often replaced by metal tubes and an original wire stock was used (or sometimes bent from thick wire).

An M3A1 cast in rubber for training purposes at Fort Lewis in Washington. The barrel and stock were made from metal. *Courtesy of Robin Bartel*

DISASSEMBLY MATS / LAYOUT CHARTS

During the post–World War II era, the "Disassembly Mat, Gun, Submachine, Cal. .45, M3A1" and the "M3A1 Layout Chart" were used. These consisted of a thick plastic foil, on which all the individual parts of the weapon were printed in black. The soldier was to place the respective parts on the correct picture when disassembling the weapon; that way it was easier to stick all the parts in the soldier's mind. The mats and charts are waterproof and easy to clean.

MOVIES

Educational movies were popular means to support the training efforts. Accompanied by morale-boosting words to the soldiers, the effectiveness and superiority of his weapons were demonstrated. A movie specifically for the M3 or M3A1 does not exist. The color movie FB (Film Bulletin) 273, ***Infantry Weapons and Their Effects***, from 1953, shows handling and operation of the following weapons: US Pistol Cal. .45, Submachine Gun Cal. .45, and US Carbine Cal. .30. The running time is eight minutes.

Members of the Special Forces get instructions in the disassembling of the M3. *National Archives*

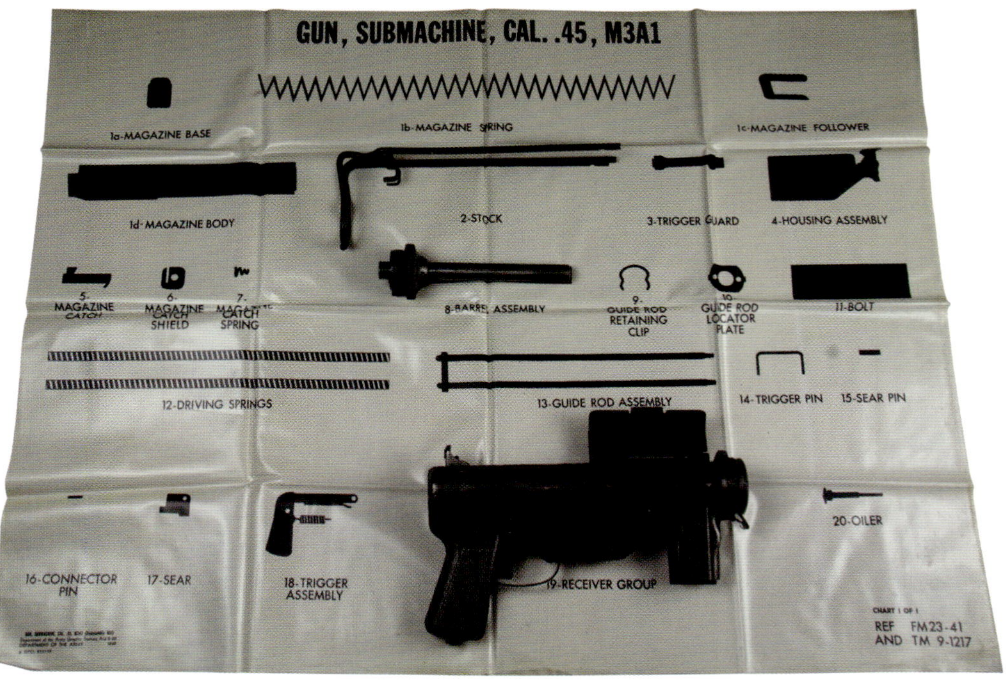

"Disassembly Mat" (Graphic Training Aid 9-60) for the M3A1, from 1949

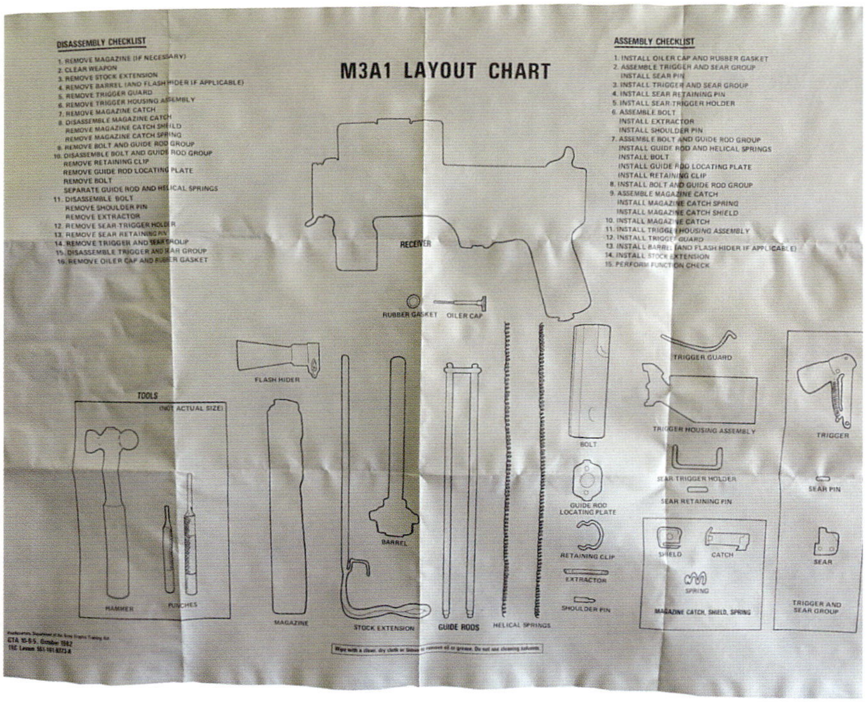

"Layout Chart" (Graphic Training Aid 10-8-5) for the M3A1, from 1982

Left: The educational movie *Infantry Weapons and Their Effects*, from 1953, shows, inter alia, how to handle the Grease Gun.

Above: Contents of a soldier's locker include an M3A1. Note the magazine pouch missing any dividers inside, so that the magazines slid around noisily when on the move.

10 AMMUNITION

AS ALREADY MENTIONED, THE Ordnance Department did not want to move away from the already adopted cartridge in .45 ACP caliber. The submachine guns in 9 mm caliber seemed to be too weak for combat conditions due to the lower man-stopping power. Other (mainly European) armies saw it differently—and so the 9 × 19 mm cartridge became the most widespread ammunition for pistols and submachine guns around the world.

The "0.45-inch Automatic Colt Pistol" (ACP) cartridge is a development of John Moses Browning for use in the Colt Model 1905 Automatic Pistol. In 1911, the cartridge was officially adopted by the Ordnance Department as the "Cartridge, Cal. .45, M1911" and was used in the M1917 revolver, the M1911 semiautomatic pistol, and all submachine guns of the US Army. Over the years, several variants, such as tracer or dummy cartridges, were put into service.

CARTRIDGE, CALIBER .45, BALL, M1911

This is the standard service cartridge of the US Army used against personnel and light materiel targets. The ball bullet consists of a metal jacket surrounding a lead alloy core, made with brass and steel cases. The Evansville Ordnance Plant (Indiana), which belongs to Chrysler, made more than 3.2 trillion cartridges in the period from 1942 to 1944. This corresponds to 96 percent of all .45-caliber ammunition produced in the States during World War II. Incidentally, the plant repaired 1,600 Sherman tanks and more than 4,000 trucks in the same period.

CARTRIDGE, CALIBER .45, DUMMY, M1921

This was used for training personnel in the operation of loading and unloading the submachine gun, and for testing weapons after repair. Different variants of these dummy cartridges have been produced in the course of time: the original M1921 had a brass case, lateral holes in the case, and a deactivated primer. For simplification purposes, the holes in the case and the primer were eliminated in 1942. From 1943 onward, the dummy rounds were made with lacquered steel cases.

CARTRIDGE, CALIBER .45, BLANK, M9

This was used to simulate fire and for salutes. The cartridge can be fired single shot only. It can be identified by the absence of a bullet and by its tapered mouth. It was adopted in November 1942.

CARTRIDGE, CALIBER .45, TRACER, M1, AND M26

This was used for observation of fire, especially at night when firing without using the sights. Secondary uses are for incendiary effect and for signaling. The bullet consists of three parts: a copper-plated or gliding metal-clad steel jacket, a slug of lead hardened with antimony, and a tracer mixture in the rear portion of the jacket. The M1 was adopted in 1939 and produced by the Frankford Arsenal and the Remington Arms Company. For easy recognition, the bullets had a red-painted tip. The bullets from Frankford Arsenal production were additionally marked with a cross-type recess on the top for identifying the tracer cartridges by touching in the dark. The first lot (FA 37) was delivered and accepted in December 1941.

Aerial view of the Evansville plant in 1944

Final inspection of cartridges. Note the high proportion of women.

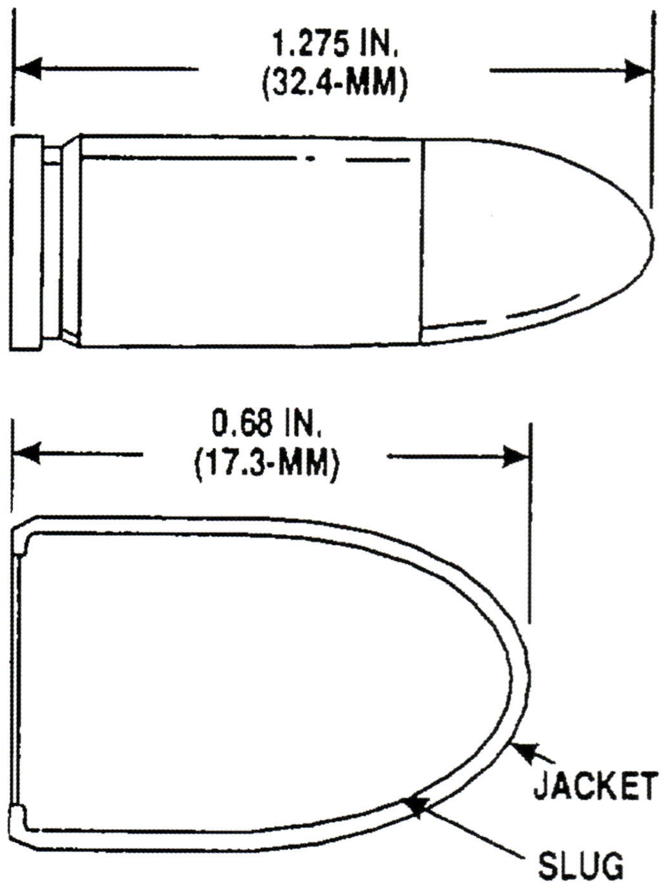

Ball, M1911, made by Evansville. *Courtesy of the Woodin Laboratory*

Dummy, M1921, made by Evansville. *Courtesy of the Woodin Laboratory*

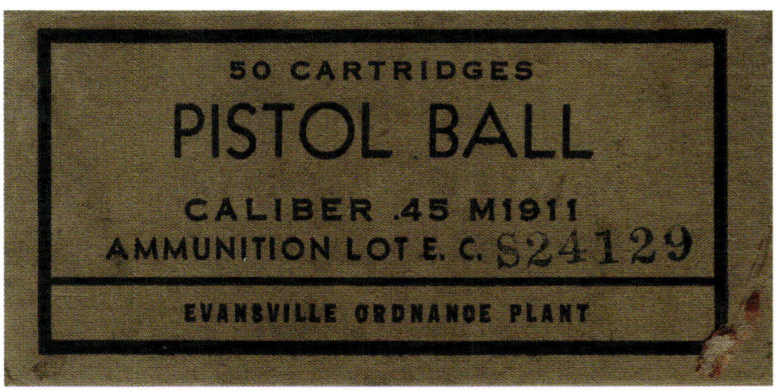

Box with fifty rounds from Evansville. *Courtesy of the Woodin Laboratory*

Twelve fifty-round boxes were packed in an airtight metal can.

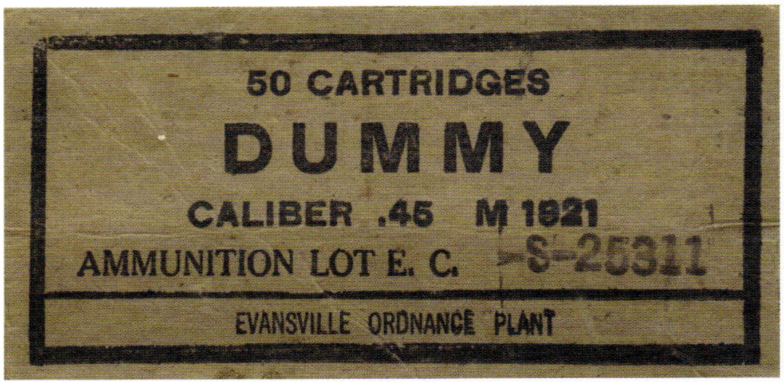

Box with fifty dummy rounds from Evansville. *Courtesy of the Woodin Laboratory*

Box with fifty blank rounds from Frankford Arsenal. *Courtesy of the Woodin Laboratory*

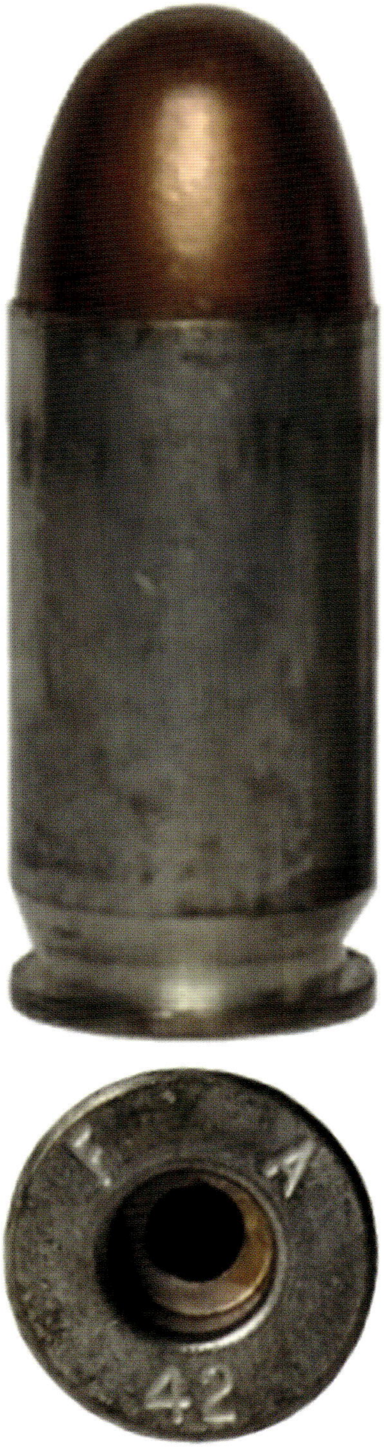

Dummy without lateral holes made by Frankford Arsenal. *Courtesy of the Woodin Laboratory*

Blank, M9, made by Evansville. *Courtesy of the Woodin Laboratory*

Initially, there were some difficulties in using the rounds with the Thompson submachine guns. Under certain climatic conditions, the tracer chemicals were inclined to swell. The pressure was so strong that the brass cases tore open from the mouth. Therefore, Frankford Arsenal developed a method (high annealing process) to make the brass cases more elastic. To identify these new cases, a star was added to the headstamp.

In mid-1945, the M1 was withdrawn from service. Further experiments had led to the tracer cartridge T30, which was then adopted in 1946 as the M26.

Tracer, M1, made by Frankford Arsenal. *Courtesy of the Woodin Laboratory*

Under certain climatic conditions, the tracer chemicals were inclined to swell. *Courtesy of RichB / International Ammunition Association*

Improved cases can be identified by the additional star in the headstamp. Here is a twenty-round box from Frankford Arsenal. *Courtesy of Guy Hildebrand*

Tracer rounds were marked with a cross-type recess on the top for identifying tracer cartridges by touching them in the dark. *Courtesy of Michael J. Tahirak*

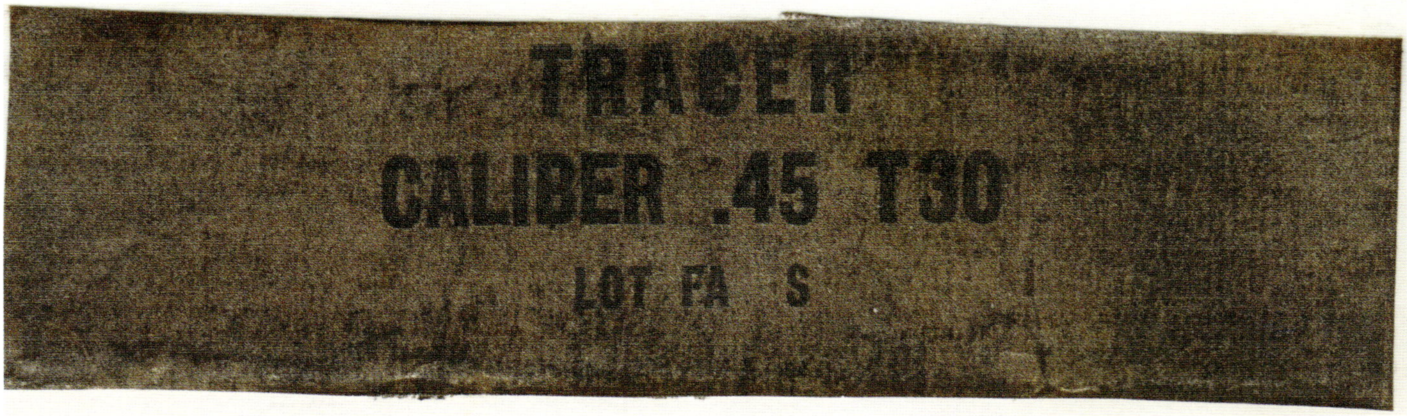

Box for the new T30 tracer round, which was adopted in 1946 as the M26. *Courtesy of the Woodin Laboratory*

Box with fifty rounds of tracer M26 of postwar production from Olin Mathieson Chemical Corporation. *Courtesy of Guy Hildebrand*

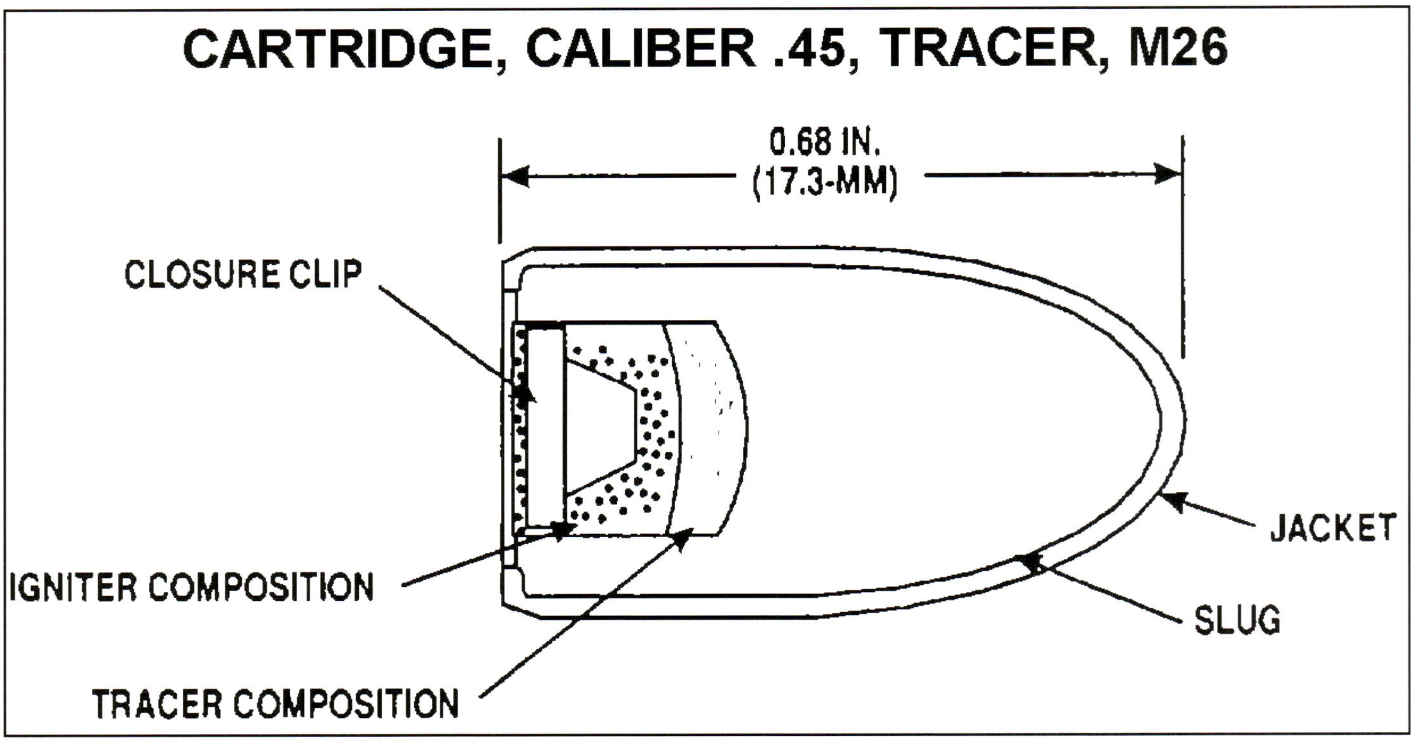

CARTRIDGE, CALIBER .45, HPT (HIGH-PRESSURE TEST), M1

This special cartridge with overpressure was used for testing of all weapons in .45 caliber before they were shipped.

High-pressure test, M1, made by Frankford Arsenal. *Courtesy of the Woodin Laboratory*

Sectional XM261 round from AAI production with a filling of sixteen spheres. *Courtesy of Paul Smith*

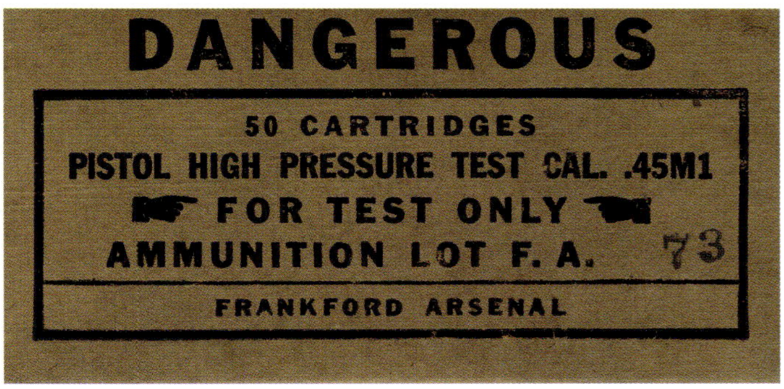

DANGEROUS

50 CARTRIDGES
PISTOL HIGH PRESSURE TEST CAL. .45M1
FOR TEST ONLY
AMMUNITION LOT F. A. 73
FRANKFORD ARSENAL

Box with fifty HPT rounds. *Courtesy of the Woodin Laboratory*

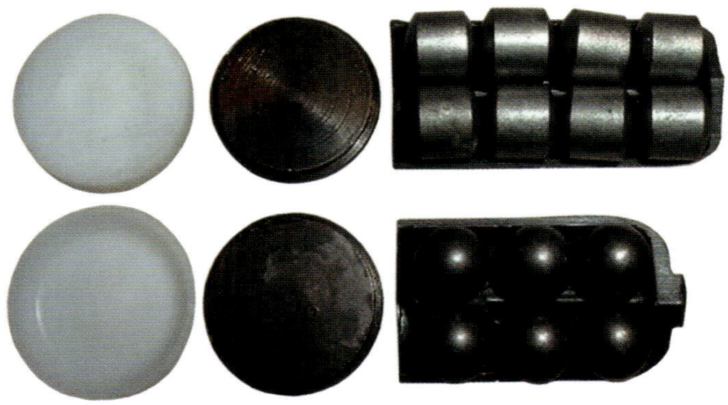

Experimental fillings for the XM261. *Courtesy of Paul Smith*

CARTRIDGE, CALIBER .45, HIGH-DENSITY SHOT, XM261

This cartridge was manufactured during the Vietnam War as part of an ENSURE (Expedited Non-Standard Urgent Requirement for Equipment) program for testing equipment under combat conditions in the hands of frontline troops. In October 1967, the AAI Corporation (formally Aircraft Armaments Inc.) was asked to make 2,000 cartridges ready for shipment to Vietnam.

A light-gray, hollow plastic bullet in the shape of the ball bullet contained a sabot filled with sixteen spheres of a tungsten alloy (Mallory 3000). Similar cartridges were tested with a filling of forty-nine flechette darts (in a 7 × 7 block).

XM261 filled with forty-nine flechette darts. *Courtesy of Paul Smith*

11 ACCESSORIES

UNLIKE MANY OTHER MILITARY weapons, the extent of accessories for the M3 and M3A1 was limited. It should just be an inexpensive and easy-to-manufacture submachine gun, which requires little maintenance and was originally not meant for special tasks. But later additional accessories followed or were at least tested, such as suppressors, curved barrels, and night visions.

OILER

For the M3, the same oiler as for the M1 carbine was used. This "Carbine Oiler, Cal. .30, M1" (5564364) was secured by a clamp made of spring steel, which was riveted to the left side of the ejector housing. The oiler was made of machined metal. It had a screw-off cap with rubber O-ring to help prevent leaks. A metal rod with tapered tip, which was fixed to the underside of the cap, oiled even the smallest parts and tight crevices. The separate oiler was dropped on the M3A1, and a larger oil can was permanently installed in the hollow grip piece.

BRUSH

Two different types of brushes were issued with the M3/M3A1: the "Chamber Cleaning Brush M6" (6108828), for cleaning the cartridge chamber, consists of a twisted metal wire handle with permanently fixed bristles, and the "Caliber .45 Cleaning Brush M5" (5,504,036), for continuous cleaning, was intended to be screwed onto the stock and to be used as a cleaning rod.

EXTRACTOR TOOL

The "Tool, Removing, Extractor" was a simple punch for driving out the extractor pin.

CARRYING SLING

Again, the already existing accessories of the M1 carbine were used. The canvas sling "Carbine Sling, Cal. .30, M1" (5557152) was now also issued with both the M3 and M3A1. And when the gray nylon sling for the carbine was adopted during the Vietnam War, it was soon used with the M3/M3A1 as well.

MAGAZINE

The magazines were not manufactured by Guide Lamp itself, but from three subcontractors.

The applied stamp "GL C-153427" refers to the number of the production drawing provided by Guide Lamp. The actual manufacturer had to add its acronym in front of this number:

KL Keeler Brass, Grand Rapids, Michigan
SPW Sparks-Withington Company, Jackson, Michigan
S unknown

Carbine Oiler, Cal. 30, M1

The oiler of the M3 was stored in a clip on the left side of the ejector housing.

The Chamber Cleaning Brush M6 with factory packaging

The Brush M5 screwed into the threatened end of the wire stock rod.

Drift punch for removing the pin of the extractor

The thirty-round stick magazine was the only one officially issued for the M3 and M3A1. Shorter magazines were made for private gun owners due to the high-capacity-magazine ban in some states in the US.

Keeler Brass on Godfrey Avenue in Grand Rapids, Michigan, 1941. Today the company is one of the leading manufacturers of furniture fittings in the United States.

The Sparks-Withington Company in Jackson, Michigan. Since 1956 the company has been named Sparton Corporation.

The manufacturer of the magazines marked with an "S" could not yet be identified.

In World War II, Bethlehem Steel was the second-largest steel producer, with more than 300,000 employees. This picture shows the grenade production facility.

View of the cannon-making facility. Bethlehem Steel could not withstand the international competition on the long term and went bankrupt in 2001.

In the records of the Ordnance Department, the Bethlehem Steel Corporation of Bethlehem, Pennsylvania, is mentioned as a magazine manufacturer, but unfortunately without listing the types of magazines. Whether this company has possibly used an "S" could not yet be clarified. Normally, however, the "S" on gun parts was used by the Seymour Products Company of Seymour, Connecticut. This company would have been experienced in the production of magazines. In addition to magazines for the Browning Automatic Rifle (BAR) and others, more than twelve million (!) magazines for the Thompsons were produced before the war ended. But the company is not listed anywhere as a manufacturer of magazines for the M3.

Despite the similar production drawings provided by Guide Lamp to the subcontractors, the magazines differ in some details, depending on the manufacturer. Magazines made by SPW differ in the shaping of the back, and also the otherwise rectangular recess for the spring-loaded magazine catch is executed in a semicircular style.

Magazines without the "GL" letters and with drawing number "C-5653427" are postwar production. They were officially purchased under the Federal Stock Number (FSN) "1005-565-3427," and then, from 1974 onward, under the National Stock Number (NSN) "1005-00-565-3427" (NATO Stock Number). Digits "5" and "6" are the country identification number (NATO country code) of the NATO Code for the National Codification Bureau (NCB); the country identification number "00" stands for the United States. Phosphated magazines were not used until after World War II.

In the technical manual TM 9-1005-229-12, from November 1969, new magazines with welded floor plates are mentioned: "Operator's repairs to the weapon will be limited to the replacement of the magazine. The magazine can be disassembled for cleaning purpose. Note: On the new magazines the base of the magazine is spot-welded to the body and disassembly of the magazine by the operator for cleaning purposes is not authorized." So far, none of these modified magazines have been found, and all later printed manuals no longer contained this note. Presumably, this kind of magazine was never issued.

Undated advertisement of the Package Machinery Company from Springfield

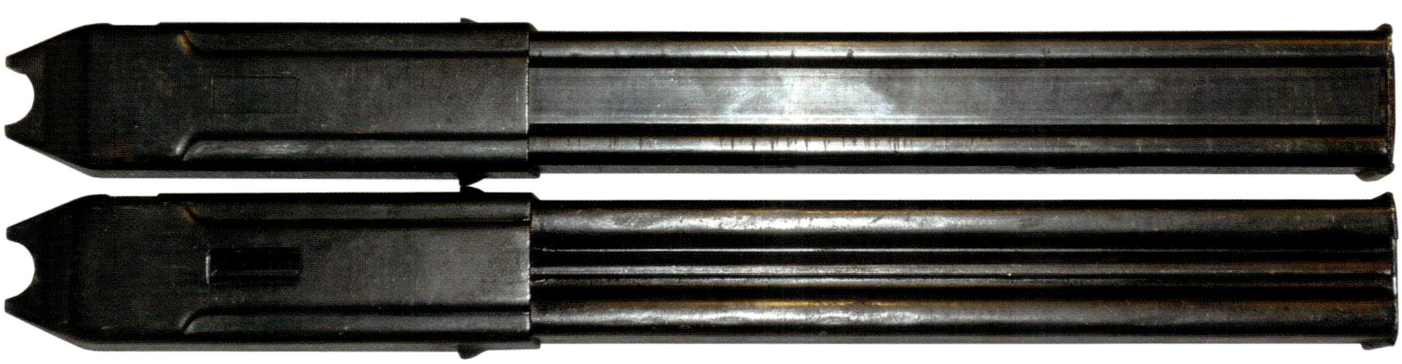

Despite the production drawing summary distributed from Guide Lamp, the magazines differ in some details. Note the different shapes of the housings: Sparks-Withington (*top*) and Keeler Brass (*bottom*).

Postwar packaging "Marvelseal" of the Marvellum Company from Holyoke

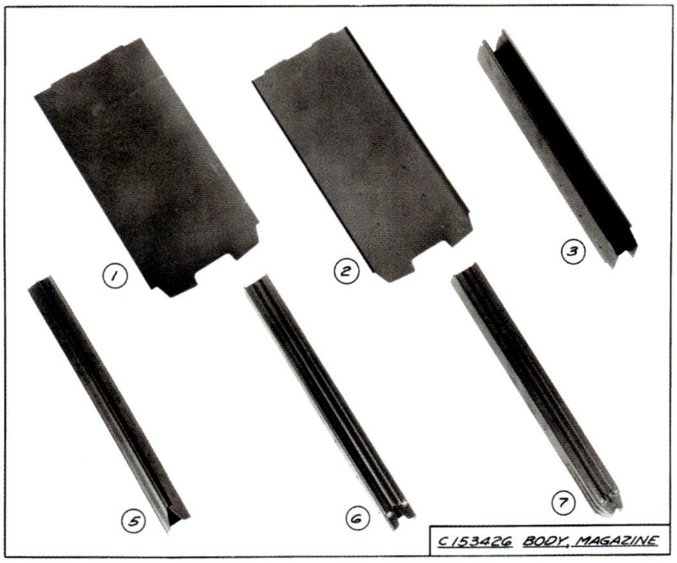

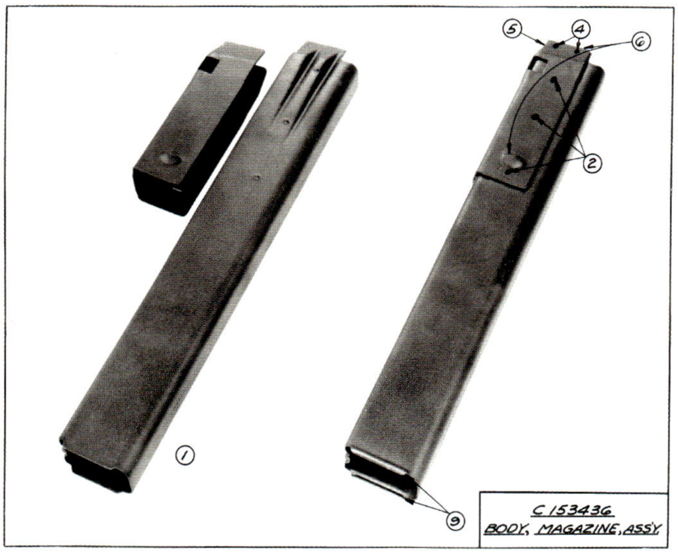

Working steps for making the magazine. *Courtesy of the Rock Island Arsenal Museum*

MAGAZINE PROTECTION CAPS

Even though every organization involved in the test period criticized the reliability of the magazines, nothing was done to rectify this problem. The magazines remained prone to dust and dirt.

Instead, a protective cap was developed by Inland together with the Eastman Company to put on the open side of the magazine. The cap was made of the transparent and hard-plastic Tenite. It remains dimensionally stable at temperatures ranging from −45°C to 115°C, has a high UV resistance, and does not react with salt water and cleaning agents, and its water absorption is very slow compared to other plastics. The cap had a formed lip on it to fit in the magazine catch recess, to hold it in place. A test at Aberdeen on June 29, 1944, ended satisfactorily. Later in the year, on November 23, the cap was officially adopted as "Caps, Magazine, T2" (7161748). The caps were intended to be disposable, mainly because they were in most cases severely damaged when removed from the magazine.

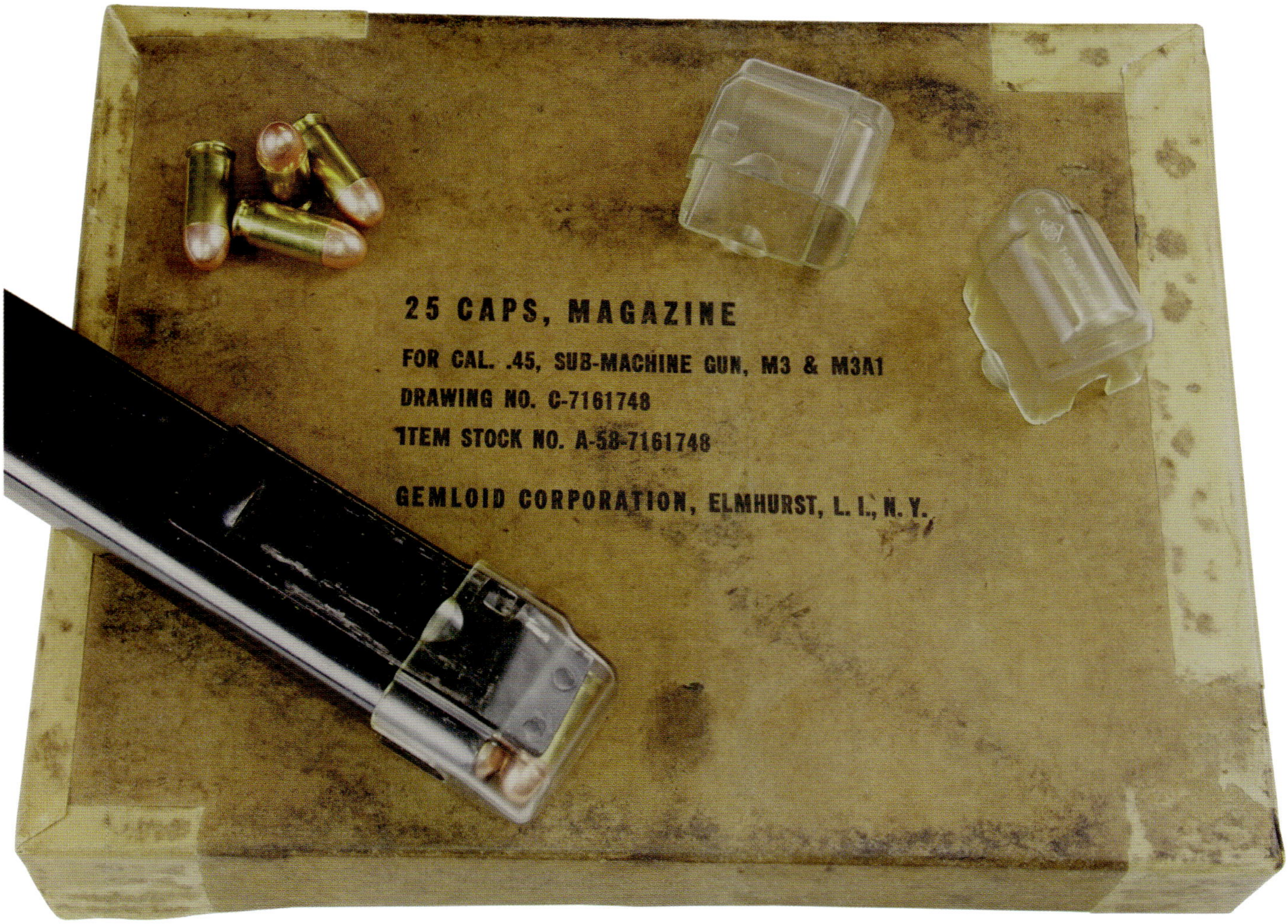

Cardboard box with twenty-five transparent protection caps made by the Gemloid Corporation from Elmhurst

The inelastic "Tenite" caps were often damaged when removed from the magazines.

The caps proved themselves in use; they protected the loaded magazines from all kinds of dirt. A disadvantage, however, was the troublesome removal of the cap: it was impossible with bare hands. Only a forceful strike against a hard object or chipping it off with a knife or bayonet helped. In the 1960s, the hard-plastic caps were replaced by a model made of soft black rubber material (7265879). They could easily—and especially noiselessly—be removed with bare hands. However, in the tropical climate of Vietnam, it turned out that the magazines and cartridges covered by the rubber cap began to rust and corrode faster than under the hard-plastic cap.

Left: The black caps from soft rubber material could be easily removed without any noise.

Bottom: Markings on a bulk pack of black rubber caps made in June 1967

MAGAZINE LOADER

Since it was almost impossible to completely fill the M3 magazine by hand, a special "Magazine Loader M1," or, according to later manuals, "Magazine Filler M1," was issued.

Magazine Loader M1

It was a very simple box-style design, made of formed sheet metal with no moving parts. With the adoption of the M3A1, the tool was no longer necessary. But although the stock contained an integrated loading tool (the bent tab welded to the underside of the wire stock), many soldiers preferred the handling of the separate magazine loader and kept it. Original loaders are unmarked and usually phosphated. In recent times, various reproductions appeared on the market, such as from Cobray for the MAC-10.

With adoption of the new M3A1 stock the separate magazine loader was no longer necessary, but many soldiers still kept this handy tool.

Usage of the magazine loader as shown in the TM-9-217 from 1943

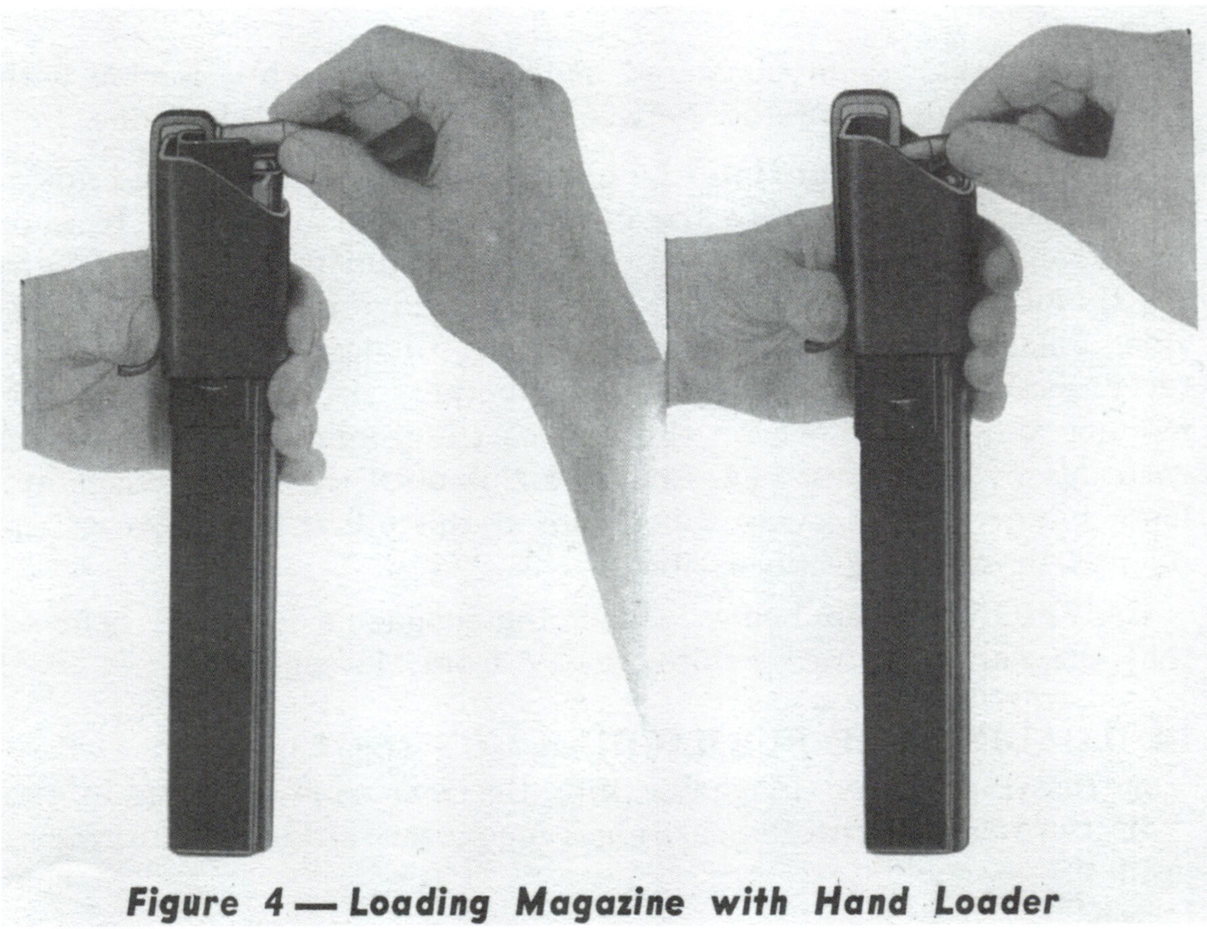

MAGAZINE POUCHES

It was originally planned to use the same magazine pouches as with the Thompson. However, this failed because the M3 magazines are about 3 cm longer and the pouches cannot be closed. So, in the time after the adoption of the M3, how to carry their magazines was up to the soldiers. On the march, there was always a space left somewhere in their equipment, but in combat, the lack of a suitable magazine pouch was very troublesome. The magazines had to be stuffed into trousers and jacket pockets, with the risk of their falling out in the movement or getting dirty easily.

A useful suggestion for combat was presented in issue 6 of ***Combat Lessons***: Three thirty-round magazines were connected with tape in such a way that one of them can always be inserted into the magazine well. So, the operator has ninety rounds for immediate use. However, the weight of three filled magazines is not to be underestimated. Often, therefore, only two magazines were taped together.

American soldiers in France. Note the stick magazines carried in the trouser pockets because there were no magazine pouches available. *National Archives*

Suggestion for increasing firepower in combat, as published in the bulletin *Combat Lessons* from 1945

Arranged fighting scene in Germany. Note that the two magazines of the soldier in the prone position are taped together. *National Archives*

While the infantry still had to struggle with its magazines, armored units received a matching pouch in late 1944: the "Case, Cal. .45 Submachine Gun Clip" (D90242) could hold eight magazines. But there were no dividers in the bag yet. So if the soldier put fewer than eight magazines in the pouch, they slid around noisily when he was on the move.

Also, the submachine gunners of this marching battle group had taped two magazines together. *National Archives*

Case D90242, as issued to the armored forces, could hold eight magazines.

The shoulder bag "Case, Ammunition, 7052438" was not assigned to a specific weapon. It was issued for carrying ammunition and magazines of various types.

TM 9-759

9-10 TANK, MEDIUM, M4A3

Item	Number Carried	Where Carried
OILER, filling, oil buffer	1	Spare parts box
ROD, jointed, cleaning, M7	1	Spare parts box
d. Caliber .45 Submachine Gun M3.		
(1) AMMUNITION.		
CAL. .45, rounds (in 30 rd. clips D35506)	900	In cases D90242
(2) EQUIPMENT.		
CASE, cal. .45 submachine gun clip	5	2 in turret under radio, 3 in right front spare parts box
e. 2-Inch Mortar M3.		
(1) AMMUNITION.		
BOMB, smoke, 2-in.	12	3 in armored box in turret 9 in armored box in left sponson over battery box
(2) EQUIPMENT.		
BRUSH, bore-cleaning, 2-in. bomb thrower, Mk 1	1	Spare parts box
CATCH, hammer	1	Spare parts box
PIN, firing	1	Spare parts box
PIN, swivel, also catch	1	Spare parts box
SPRING, catch	1	Spare parts box
SPRING, catch, junction nut	1	Spare parts box
SPRING, main	1	Spare parts box
SWIVEL, hammer	1	Spare parts box
f. Hand Grenades.		
FRAGMENTATION, Mk 11	4	In turret
SMOKE, WP M15	4	In turret
SMOKE, colored, M16 or M18	4	In turret

Inventory list of the Medium Tank M4A3 Sherman: thirty magazines (900 rounds) in five cases D90242

A soldier of the 4th US Infantry Division with a shoulder bag for carrying magazines. The picture was taken on July 24, 1944, the day before the breakout from the Normandy bridgehead, known as Operation Cobra. *National Archives*

Battle of the 97th US Infantry Division in Siegburg on April 10, 1945. Note the bag with magazines and the leather case M1918 for spare parts and tools. *National Archives*

The "Pocket, Ammunition, Magazines, SMG M3," with three separate compartments for the magazines

Well into the 1950s (the exact date is disputed), a special magazine pouch, "Pocket, Ammunition, Magazines, SMG M3," holding three magazines, was adopted. Each magazine sat in a separate compartment with press button and could be removed individually.

CASES AND COVERS

For carrying accessories and spare parts of the M3/M3A1, various cases had been available. The dates when these items were adopted, replaced, or withdrawn from service cannot be precisely determined. The different editions of the manuals, however, give at least some rough guides.

Along with the first M3s, the "Accessories and Spare Parts Case M1918" was issued. This boxy leather case originally came from the BAR M1918 and was then also used for other weapons. For example, it can be found in the Thompson manuals. The case was even large enough to stow the magazine loader.

For smaller parts, there was the "One Button Fabric Envelope 3 × 3½ in.," made of greenish canvas.

In the 1960s, it was replaced by the somewhat larger "Two Button Fabric Envelope 4-7 / 8 × 3" (7228907).

Together with the M3A1, another accessory case was introduced. The "Spare Parts Bag" (5559738) could be closed by means of a drawstring and should properly contain the following parts:

- guide rod clip (5349931)
- washer (7161911)
- trigger guard (6301456)
- magazine catch (6301453)
- magazine catch shield (7160997)
- extractor (6301464)
- sear (7162774)
- spring (7160998)

The "Accessories and Spare Parts Case M1918," made from leather

Small spare-parts envelope with one push button

Larger spare parts envelope with two push buttons

The "Bag, Spare Parts" was adopted with the M3A1.

The "Cover Grease Gun" is often offered on the collector market as an accessory, but in fact it was a cover for the actual grease gun for motor vehicles.

It should be noted that the "Cover Grease Gun–C6943" was not meant for the submachine gun M3 or M3A1. However, it is often offered as an accessory, intentionally or unintentionally. In fact, it was a cover for the actual grease gun for motor vehicles.

FLASH HIDER

During the experiments with suppressors in March 1945, it was noticed that the muzzle flash was eliminated when such a device was being used. With this in mind, an official contract for the development of a muzzle device to reduce the muzzle flash was carried out on June 14, 1945. Shortly afterward, the first prototype was ready: a funnel-shaped flash hider made of sheet metal that had to be pushed onto the barrel and fixed by a clamp with a thumbscrew. The production of this "Hider, Flash, T-34" started before the Japanese surrender, but the device arrived too late at the front to see active service in combat.

In 1950, the production was resumed for the Korean War with the new designation "Hider, Flash, M9" (7265628). Two different types were manufactured: one with a welding seam along the body of the cone (manufacturer: "LFC" Landers, Frary & Clark) and one without a visible seam (manufacturer: "Marco U.S.A."). Both models were phosphated gray.

The flash hider M9 was used for the first time in Korea. Opinions differ as to its practical usefulness.

M3A1 with flash hider M9

Battle for the Citadel and the Imperial Palace of Hué in February 1968. Note the flash hider and the unusual magazine pouch.

Comparison of the flash hiders M9 made by MARCO (*left*) and LFC (*right*)

Packaging of a flash hider made by LFC

CANVAS PROTECTIVE COVER

The "Cover Cal. 45 SUB" (d90196), made from canvas, was to protect the submachine gun from dirt and damage. It was used mainly in vehicles in which the weapons were carried for emergencies and had to be within easy reach. The shape of the cover makes it possible to stow the weapon with a standard thirty-round magazine inserted.

WATERPROOF PROTECTIVE COVER

For amphibious operations, waterproof protective covers made from the plastic material Pliofilm were available for weapons and equipment, in different sizes.

The "Cover, Waterproof, Submachine Gun" (Stock No. 74-C-310-45) was intended for the Thompson models and the M3/M3A1. Pliofilm is a rubber-based, yellowish (and later greenish), transparent plastic that was developed in 1934 by the Goodyear Tire & Rubber Company.

GUN RACKS

To store weapons in barracks, strongholds, or perhaps vessels, various racks or wall brackets were used. Some, such as the "Small Arms Storage Rack M7" (8407182), could be obtained via the official channels. Others were made by the units themselves.

TRANSPORT AND STORAGE CONTAINERS

Normally, the arms were factory-packed in wooden crates and delivered to the recipients. For temporary intermediate storage, they remained mostly in those boxes.

For airborne troops and to supply ground forces from the air, various containers were available. The most common one was the "A-5 Aerial Delivery Container" for airdrop, with space for twelve submachine guns with accessories.

To supply the resistance fighters in Europe, a container of British origin was used. The M3 and M3A1 (some of them converted to 9 mm caliber), however, were seldom used for that purpose. The bulk of the aid deliveries consisted of 9 mm Sten guns. And of these weapons, only a small part reached the resistance fighters, because many containers landed on the German side or could not be recovered quickly enough.

In 1948, for long-term storage of weapons under different climatic conditions, the barrel-shaped "Longtime Storage Carrier" was developed. It took ten M3s or M3A1s with magazine and sling. Two bags with desiccant were placed on both sides inside the container, and the rims were equipped with washers and sealed up airtight. The insulation contains asbestos and is therefore hazardous to health from today's perspective. Therefore, the display opening of the exhibit in Springfield Armory (see illustration) is closed with a Plexiglas plate and carefully sealed.

9 MM CONVERSION KIT

One of the most desirable accessories for collectors is the "9 mm conversion kit." It was developed at the suggestion of the Office of Strategic Services (OSS) and was originally intended to adapt the weapon for the European theater. The British used 9 × 19 rounds, as did the German army, so the resistance fighters would have been easier to supply with this kind of ammunition instead of the .45 ACP rounds. However, since the British preferred their own Stens, only very few conversion kits actually found their way to Europe.

The conversion kit consists of three parts:

- C153425 bolt assembly
- C153435 barrel assembly
- C153430 adapter

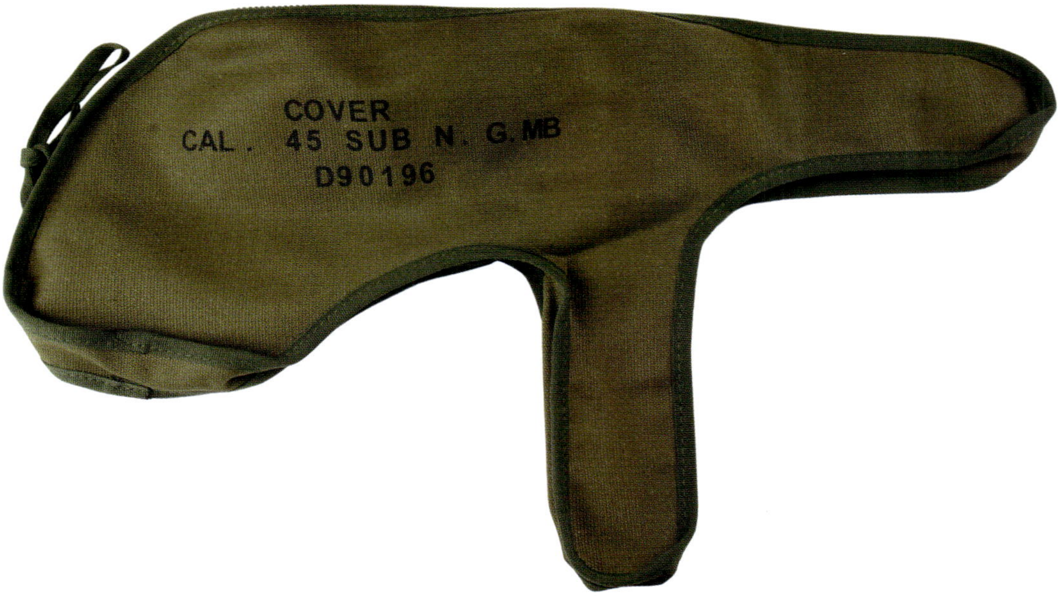

The "Cover Cal. 45 SUB" was used to store an M3/M3A1 with magazine ready for use in vehicles.

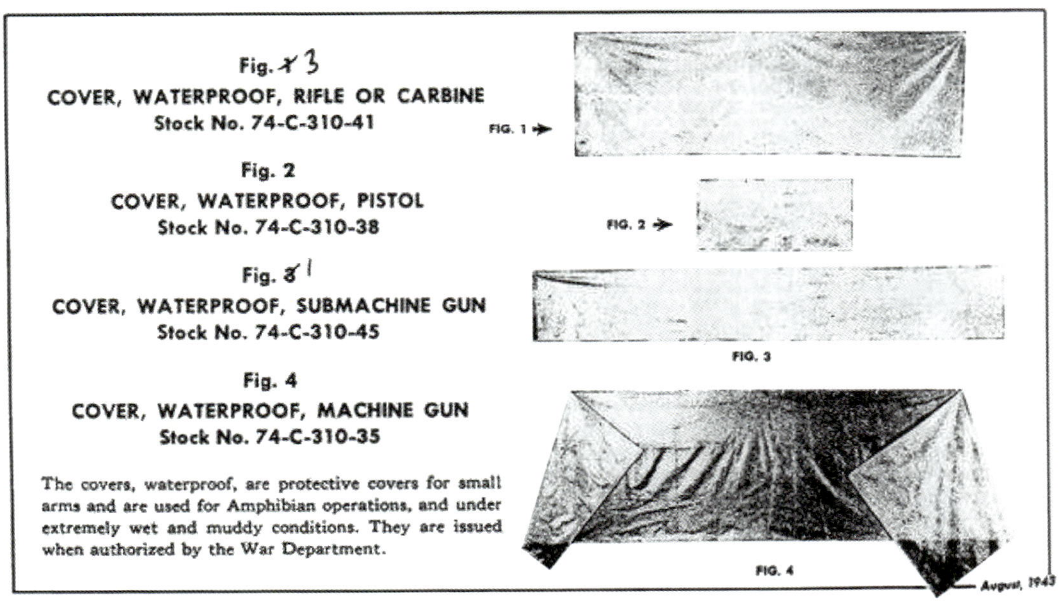

Fig. 3
COVER, WATERPROOF, RIFLE OR CARBINE
Stock No. 74-C-310-41

Fig. 2
COVER, WATERPROOF, PISTOL
Stock No. 74-C-310-38

Fig. 1
COVER, WATERPROOF, SUBMACHINE GUN
Stock No. 74-C-310-45

Fig. 4
COVER, WATERPROOF, MACHINE GUN
Stock No. 74-C-310-35

The covers, waterproof, are protective covers for small arms and are used for Amphibian operations, and under extremely wet and muddy conditions. They are issued when authorized by the War Department.

For amphibious operations, waterproof protective covers were issued (illustration from QM 3-1, dated 1943).

The "Cover, Waterproof, Submachine Gun" for the Thompson, M3 and M3A1

The "Storage Rack M7" for twenty weapons in two rows

Self-made rack for five weapons from a military facility. *Courtesy John D. Bush*

The "Longtime Storage Carrier" with space for ten M3s or M3A1s. *Courtesy of the Springfield Armory NHS / US NPS*

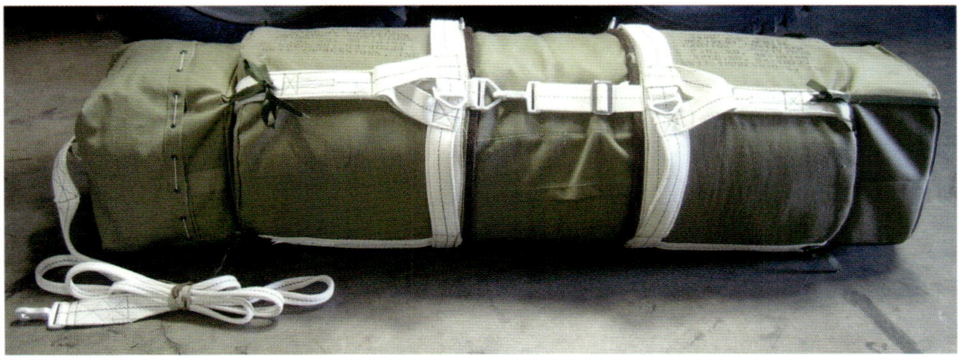

"A-5 Aerial Delivery Container" for airdrop, with space for twelve submachine guns with accessories. *Courtesy of Robert L. Marvel*

British airdrop container to supply resistance fighters in Europe. Many of these containers landed on the German side or could not be recovered quickly enough.

Experimental M3E1 with serial number X41. This weapon was used to test the conversion kit. Note the marking "9 MM." *Courtesy of the Springfield Armory NHS / US NPS*

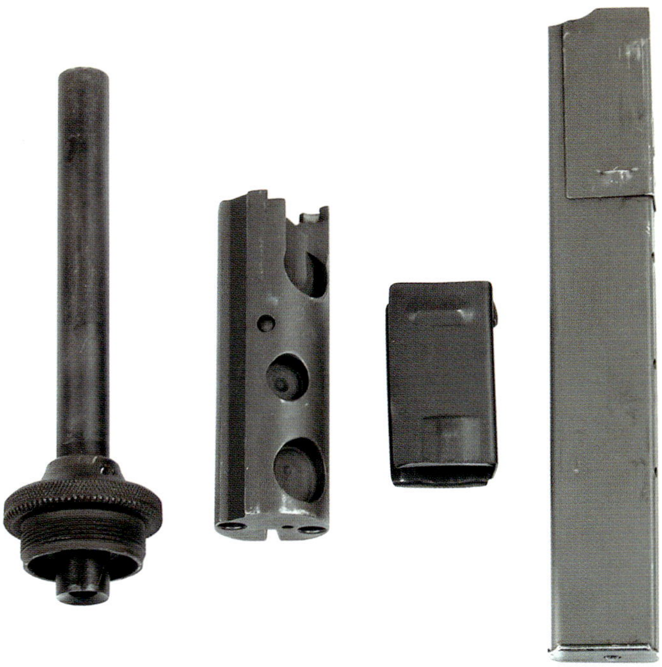

The "9 mm Conversion Kit" for the M3A1 consisted of barrel, bolt, and adapter. The magazine is from the Sten. *Courtesy of Frank Iannamico*

A 9 mm bolt installed in an M3A1. *Courtesy of Frank Iannamico*

A 9 mm bolt for the M3A1 made by Rock Island Arsenal (RIA). *Courtesy of Frank Iannamico*

The magazines of the Sten submachine gun were obtained from the British government.

In December 1943, the Ordnance Department suggested a small series of 500 conversion kits for the M3, which was approved in February 1944. The bolts were made by Buffalo Arms, New York; all other parts were made by the Rock Island Arsenal, Illinois.

The magazines were purchased from Britain. Later, also conversion kits for the M3A1 with an adapted bolt were manufactured by the Rock Island Arsenal. The exact production quantities of both types are not known, but most of the still-existing conversion kits contain the M3 bolt.

The bolt face had to be modified because of the longer 9 mm barrel. Picture taken in April 1944. *APG*

Comparison of the two barrels: .45 caliber (*left*) and 9 mm (*right*). *Courtesy of Frank Iannamico*

The adapter C153430 for the magazine well to hold the slightly smaller Sten magazines. *Courtesy of Frank Iannamico*

Converting an M3 or M3A1 is done very quickly: barrel and bolt are exchanged and the adapter, made of thin spring steel, is inserted into the magazine well. The adapter has two formed lips on its sides that fit in two recesses inside the well to hold it firmly in place. However, the removal of the inserted adapter is tricky. It is easier if the magazine release button is removed beforehand.

CURVED BARREL

In 1942, the German Reich made its first attempts to develop a ***Krummlauf*** (curved barrel) for the standard Karabiner 98k (7.92 × 57 mm caliber). Later the device was modified for use with the new revolutionary assault rifle Sturmgewehr 44 (short cartridge, 7.92 × 33 mm caliber). With this curved barrel, the soldier could shoot from cover, without being seen by the enemy. When installed in armored vehicles, the defense against enemy infantry in the otherwise blind spots would have been possible. The tests were satisfactory, but the development could not be completed before the war ended. None of these devices were officially adopted, and only a few prototypes were made. Most of them were captured by the Allies and taken home for further examination.

Inspired by the German curved barrels, similar experiments were carried out in various countries after the war. The US Ordnance Department appeared enthusiastic about the idea, especially as armament for armored vehicles against approaching enemy infantry.

First, a slightly longer barrel was bent to 30 degrees and inserted into the standard barrel nut. With this angle, the soldier would be able to shoot from safe cover. But because of the absence of an optical sighting device (periscope), precision shooting was impossible. The project remained in the test status.

This Italian partisan is equipped with an M3 in 9 mm caliber. Note the Sten magazine in the weapon and the magazine pouch.

Italian partisans in the city of Cuneo, in Piemont. Their armament is a mixture of at least five American M3s, two British Bren guns, two German MG42s, and one Italian MG Breda 30. The young boy in the foreground is equipped with a Finnish Suomi M/31 submachine gun.

A group of Italian partisans are busy with the exchange of messages. One of them carries an M3.

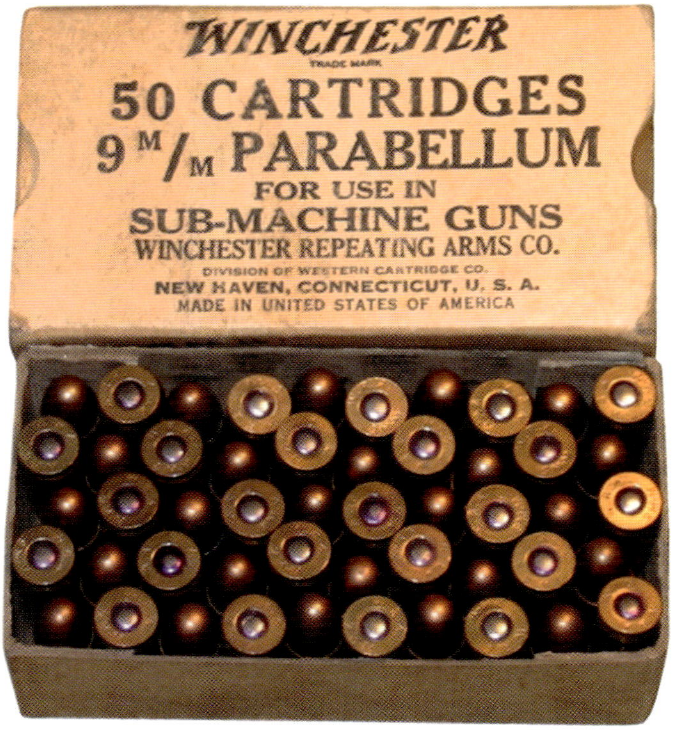

Package of 9 mm cartridges made by Winchester especially for use in submachine guns

The next development step was a 90-degree curved barrel for installation in tanks. But this time, the barrel was not bent, but instead an extension piece was added. This extension was pushed onto the standard barrel and fixed to it by means of a freely rotatable clamp piece. The curved part of the extension was half-cut open to the inside of the bend and thus formed a channel in which the projectile could travel along. The opening prevented projectiles getting stuck in the bend. When installed in a mount in the sidewall or roof of a tank, the weapon could be pivoted by 360 degrees, thanks to the rotatable clamp. The horizontal and vertical pivot ranges were 40 degrees. At the end of 1945, tests were carried out with an M4 Sherman tank, but the result was very unsatisfactory. The barrel was not resistant enough against the torsion forces during firing, and the hits were spread extensively over the target. The chance of a well-aimed hit was very low. These problems were the same as the German engineers already had to defeat, and therefore their tank version, called "extension P" (P = Panzer), with ball mount and periscope, became very robust and heavy.

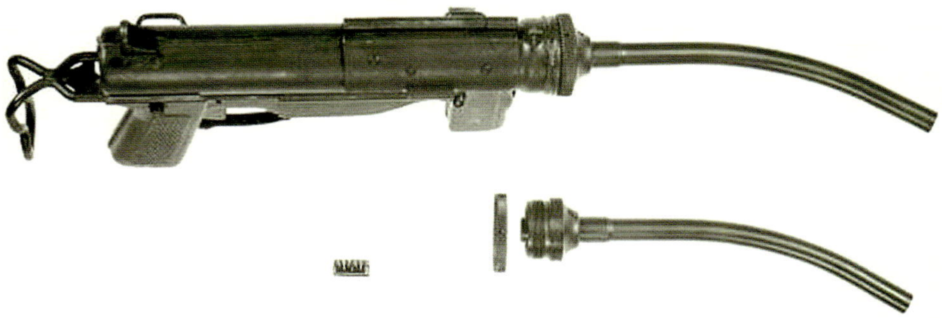

Springfield photo of an M3A1 with curved barrel (30 degrees) from February 1947. *Courtesy of the Springfield Armory NHS / US NPS*

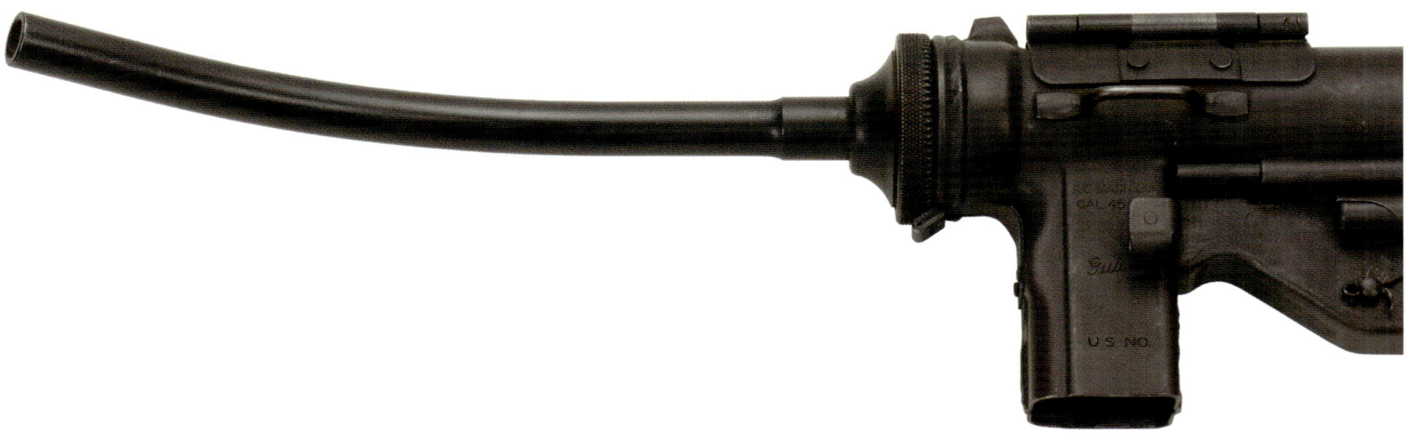

This unnumbered M3 was used to test the curved barrel device with 30 degrees of bending. *Courtesy of the Springfield Armory NHS / US NPS*

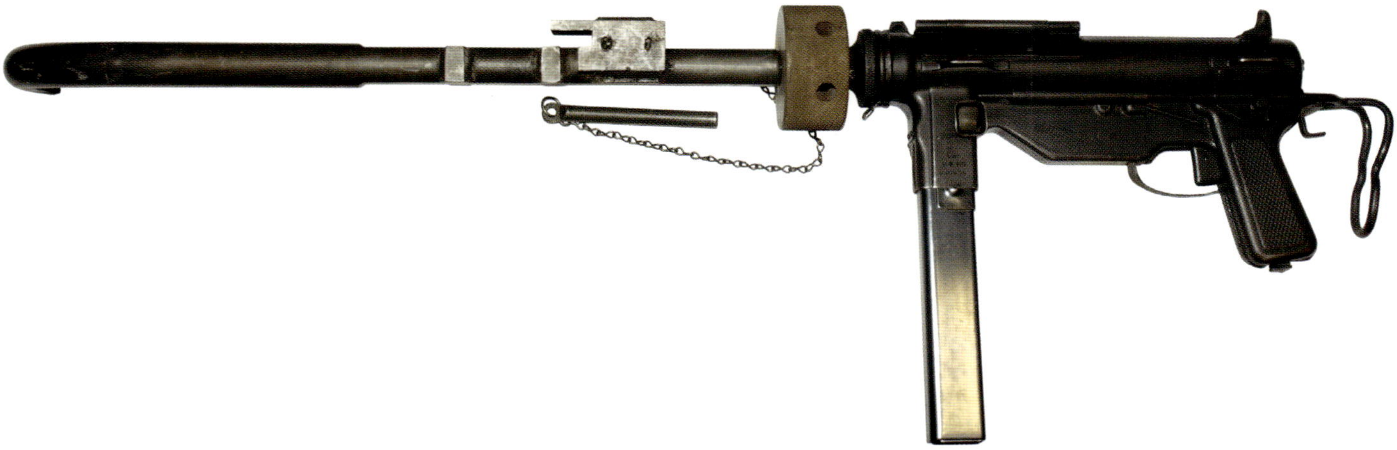

M3A1 with a 90-degree curved barrel device for installation in armored vehicles. *Courtesy of the National Museum of the Marine Corps, Quantico*

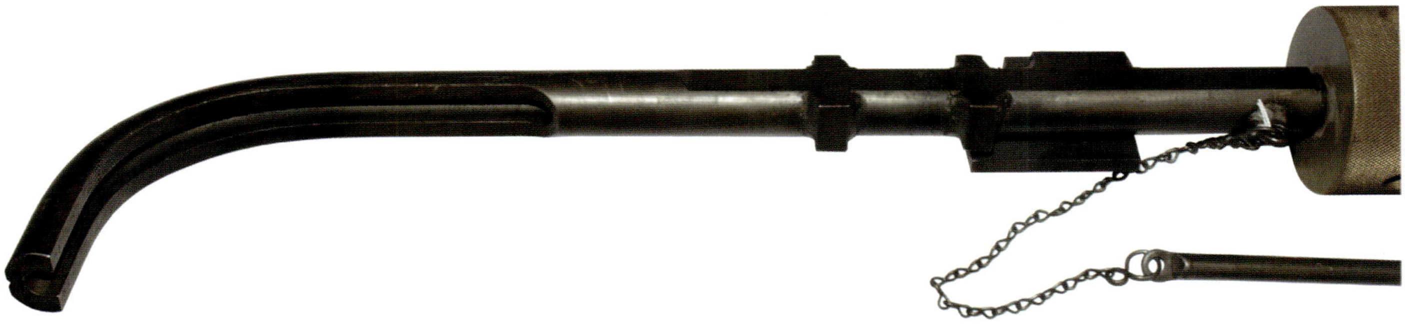

The bent part of the barrel was cut open to prevent the bullets from getting stuck. *Courtesy of the National Museum of the Marine Corps, Quantico*

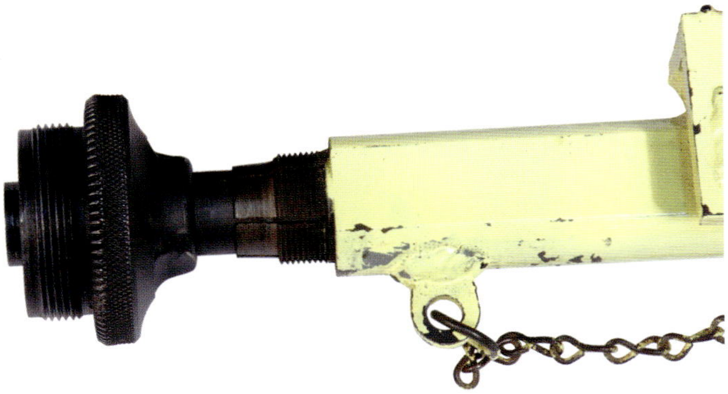

The curved barrel device was pushed over the standard barrel and fixed by a clamping nut. Note the multiple slotted end with thread for the nut. *Courtesy of the National Firearms Centre / Royal Armories, Leeds UK*

The chained-up stick was a tool to firmly tighten the clamping nut. *Courtesy of the National Firearms Centre / Royal Armories, Leeds UK*

Another series of experiments were conducted in Aberdeen, this time with the tank models M4A3 Sherman, M26 Pershing, and M46 Patton, and the submachine guns were fixed in the hatch of the commander and the top of the turret over the loader. But again, there was a lack of stability of the curved barrels, and without an aiming device, there was only a small benefit. The overall result was not worth the technical effort.

Press photo from January 1951: "Sgt. Donald D. Rector of Muskegon, Michigan, demonstrates how a newly developed barrel makes it possible to shoot around corners with a standard submachine gun. The barrel, which fits on the army's standard M3 'grease gun' was developed at the Detroit arsenal for use by tanks crews and infantrymen."

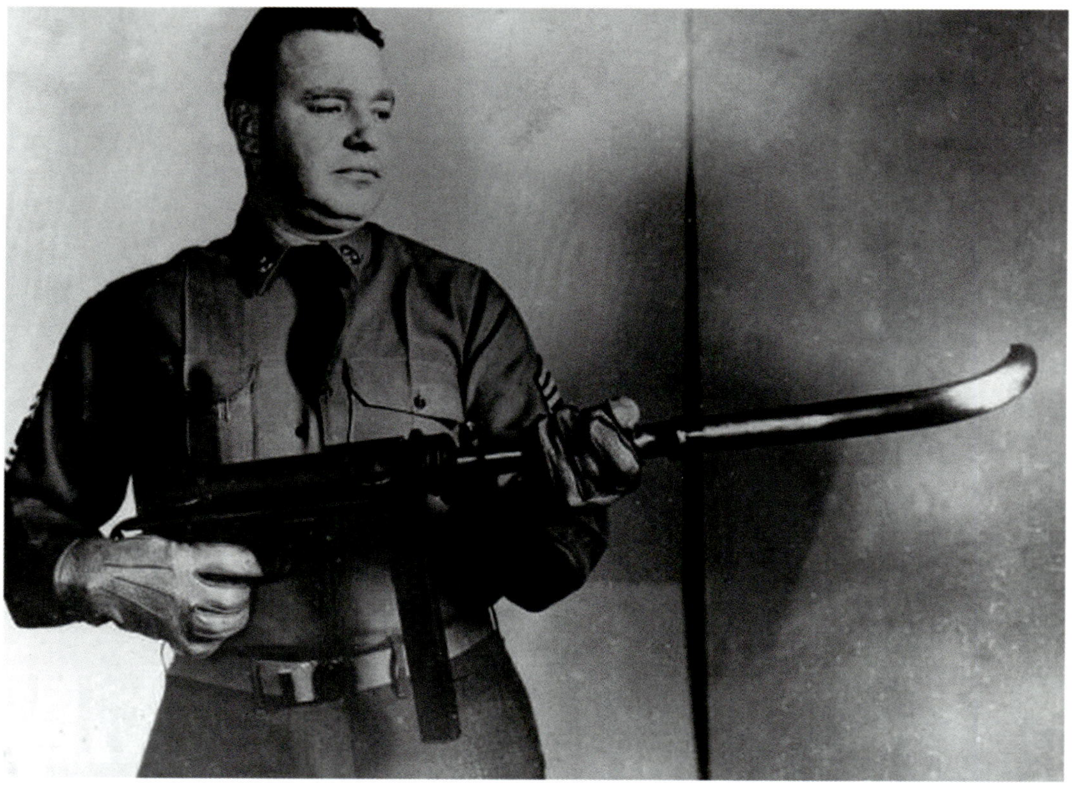

Press photo from 1953. Attached to the device is a mount for fixing it to a vehicle.

Test of an M3A1 with night vision device. The combination did not prove successful.

NIGHT VISION DEVICE

During World War II, the US Army developed night vision devices (infrared) for handguns. However, the development lasted almost to the end of the war, and so these devices did not see action before the invasion of Okinawa in April 1945. These devices were mounted on the select-fire M2 carbine, and some good success in the defense against Japanese night attacks was achieved. The bulky, heavy, and very sensitive infrared device was improved in the following years and was again used with the M2 carbine in the Korean War. A version for the M3A1 was tested but not adopted.

The M1 carbine with "Infrared Sniperscope M3" was used in combat in larger numbers. *Courtesy of the Rock Island Auction Company*

12 THE SUPPRESSED M3 AND M3A1

THE SUCCESSFUL USE OF silenced Sten Mk. II S for the British Special Operations Executive (SOE) also aroused the US Ordnance Department's interest in a suppressor for submachine guns (SMGs). Founded in June 1942, the Office of Strategic Services (OSS) could use such a weapon for its own special operations and support of resistance groups in Europe and the Pacific theater. The M3 seemed to be well suited for that purpose.

Because standard .45 ACP rounds fired from short-barreled submachine guns are inherently subsonic, it is one of the most powerful pistol calibers available for use in suppressed weapons. Suppressors reduce the audible report by slowing down and channeling the high-speed gas generated by the burning and expanding gunpowder before it exits the muzzle. Suppressors cannot act on a supersonic shock wave continuously generated by a bullet exceeding the 1,084-feet-per-second (332 meters per second) speed of sound at 32°F (0°C) ambient temperatures. This shock wave is continuously produced throughout the entire flight path over which the bullet is supersonic, which extends long after it exits the barrel. One major disadvantage of automatic weapons, however, remains: the unavoidable noise of the moving internal parts when the spent case gets extracted and the next round gets automatically loaded. So a weapon with a suppressor is not completely silent, but it lacks the typical loud report, and for the enemy, it is more difficult to locate the direction of the sound's source.

The first tests with suppressed M3 submachine guns had already been held in 1942, at Fort Benning, Georgia. These devices were based on the construction of Hiram Percy Maxim, with curved vanes of thin sheet metal, but were not convincing. The design was heavy, the gun became extremely top heavy, and the suppressor heated up quickly during continuous fire. Two different models were tested: a long one with a 20-inch barrel and a shorter one with a 12-inch barrel. Because of the unsatisfactory results, the tests were discontinued.

In 1943, the Ordnance Department contacted the Bell Laboratories Company on behalf of the OSS concerning the development of a suppressor for the M3. The production of the suppressor parts and the assembly took place at the High Standard Manufacturing Company in Hamden, Connecticut. The production of special barrels and modified barrel nuts (collars) was assigned to Guide Lamp, and the first order was for 1,000 devices. Whether and in what quantity follow-up orders had been placed is not known. One of the big advantages of this suppressor design was that the standard M3 barrel could be removed and the suppressed barrel could be quickly screwed on in its place.

The outer form of the special barrels of Guide Lamp did not differ from the standard serial barrels. However, a total of forty-eight holes were drilled into the barrel (in four rows of twelve holes) to allow gases to escape as the bullet was fired. Each hole had a diameter of ¼ inch. The muzzle was threaded on a length of about ½ inch. The barrel nuts were taken from the serial production and were modified at the front to serve as a support for the rear end of the suppressor.

Suppressed Sten Mk. II S as used by the British SOE. The US OSS was impressed by its performance. *Courtesy of the Military History Institute & Army Museum Prague*

M3A1 (serial number 675550) with the suppressor designed by Bell Laboratories. *Courtesy of the Springfield Armory NHS / US NPS*

The suppressor could be attached to both the M3 and the M3A1. *Courtesy of the Springfield Armory NHS / US NPS*

The barrel nuts have been modified at the front to serve as a support for the rear end of the suppressor. *Courtesy of the Springfield Armory NHS / US NPS*

The Bell suppressor taken apart. Note the stack of 230 perforated discs of steel wire mesh. *Courtesy of Corey A. Shaffer / Thompson Machine Company*

The standard M3 barrel could be removed and the suppressed barrel could be quickly screwed on in its place. *Courtesy of the Springfield Armory NHS / US NPS*

Suppressor collection of a US agency. *Courtesy of Corey A. Shaffer / Thompson Machine Company*

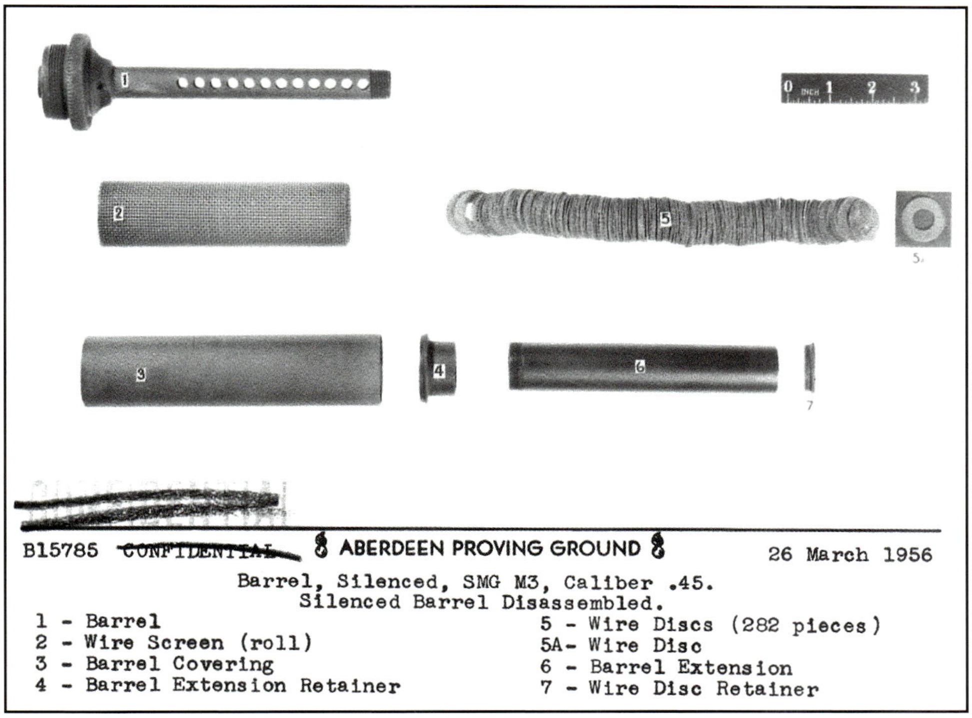

The suppressor taken apart: *1.* barrel; *2.* wire screen (cylinder of stainless-steel mesh); *3.* barrel covering (expansion chamber); *4.* barrel extension retainer (connecting piece); *5.* wire discs (perforated stainless-steel mesh); *6.* barrel extension; *7.* wire disc retainer (front end plate). *APG*

The suppressor itself comprises two tubes. The rear tube (barrel covering), 7½ inches in length and 1½ inches in diameter, forms the expansion chamber and is filled with a stainless-steel wire mesh cylinder, which is fitted over the barrel. The threaded muzzle part of the barrel protrudes from the expansion chamber. The actual suppressor housing (barrel extension) is attached to the barrel covering by a connecting piece (barrel extension retainer) that is screwed onto the muzzle thread of the barrel, thus locking the covering in place against the modified barrel nut. The extension, 7 inches in length and 1.13 inches in diameter, contains 230 perforated discs of stainless-steel wire mesh. The front end of the extension is closed by a threaded end plate.

When the projectile travels down the bore, some of the propellant gases escape through the forty-eight holes in the barrel into the expansion chamber, where they are slowed down and cooled by the steel wire mesh. When passing through the connecting piece, the projectile leaves the barrel and moves through the perforated steel wire mesh discs in the barrel extension. Here the remaining gases are trapped for a short time and cooled down and then escape into the surrounding air.

The influence of the suppressor on the noise level is described in a test report from Aberdeen Proving Ground as follows: "While the silenced barrel does not completely eliminate the sound, the reduction of the report makes it difficult to determine the exact location of the weapon when being fired at a distance of fifty yards or greater. The sound heard by the individual varied with his position with respect to the weapon. When the individual was in the line of fire or forward of the gun muzzle, a louder sound was heard than when positioned at the rear of the gun. When positioned at the side or at the rear of the gun the noise caused by the operation of the mechanism was most noticeable."

The use of the suppressor, however, changed the point of impact of the projectile on the target. Tests showed that an M1 steel helmet will not be penetrated at a distance of 75 feet. Using a standard barrel without the device, it was no problem to perforate the helmet even at greater distances. Also, it was found that the projectiles sometimes struck the edge of the central opening in the front-end plate of the barrel extension, thus damaging it. Pieces of the bullet jacket were lodging in the sleeve behind the plate. In some cases, the end plate was shot out of the sleeve, and the installed wire discs fell out. This problem was caused by the not-completely-rigid connection of the two sleeves. The connecting piece is only screwed onto the muzzle thread, and so the alignment of the barrel extension with the bore of the barrel is difficult because the suppressor is too flexible. The connection does not withstand the leverage effect of the long protruding extension. A slight enlargement of the opening in the plate to a 0.550-inch diameter solved the problem.

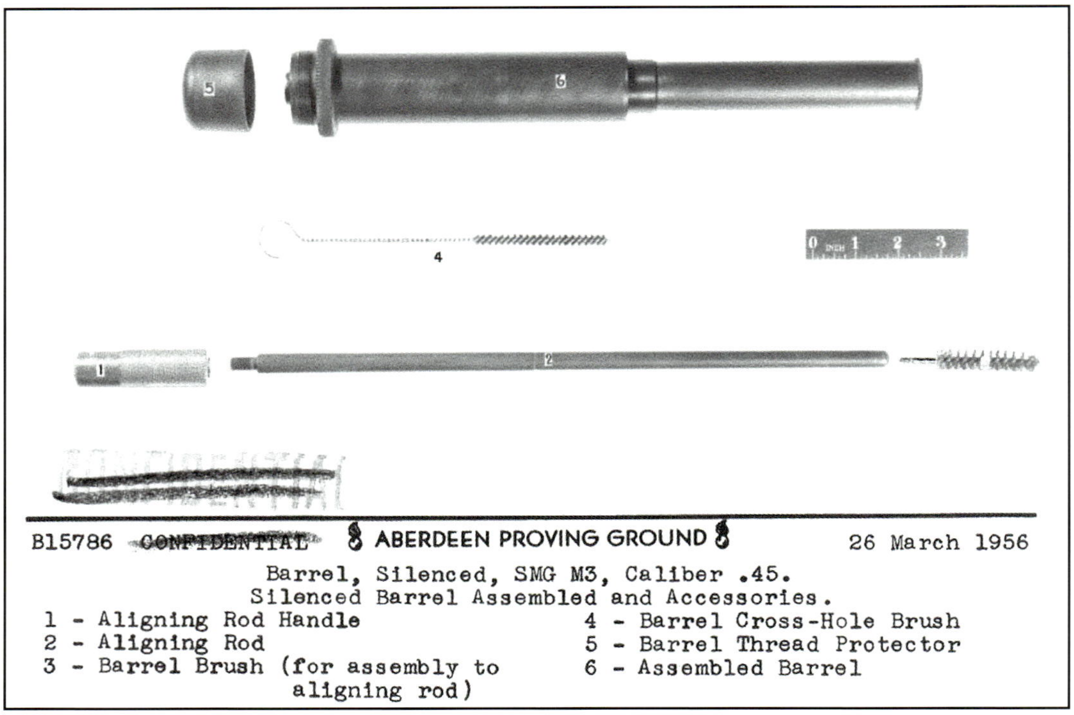

Accessories: *1.* aligning rod handle; *2.* aligning rod; *3.* barrel brush; *4.* barrel cross-hole brush; *5.* barrel thread protector; *6.* assembled barrel with suppressor. *APG*

The care and maintenance of the suppressor proved to be somewhat complicated. In particular, the perforated steel wire mesh discs were difficult to retain in the barrel extension. If they slipped in the housing or became deformed, the path of the projectile was obstructed, and it could hit and damage the discs (thus reducing the suppressing effect). As already mentioned, the barrel extension could deviate from the exact alignment with the barrel easily—especially during transport. For this reason, a special tool was issued with each suppressor: An "aligning rod" should be installed inside the suppressor during transport or shipment as a stabilizer. It reaches from the cartridge chamber of the barrel to the front-end plate of the barrel extension and should prevent bending of the two sleeves. By means of a detachable handle, this stabilizer could also be used as a cleaning rod. Every suppressor was also issued with a special "barrel cross-hole brush" for cleaning the forty-eight small, lateral holes, and a sheet-metal "barrel thread protector" to be screwed on the thread of either the suppressor or barrel nut (depending on which part was in use). Two versions of the protector cap were made. The early one with a smooth back was often hard to remove, so the design was changed, and the later ones got a square pressed-in recess for applying a tool.

According to the reports, the use of the suppressor improved the accuracy when firing automatically. This fact is probably due to the increase in barrel weight and dimensions and to the lower velocity, causing lower recoil energy. Another positive effect is the elimination of the muzzle flash. No flash was observed when the suppressed weapon was fired at night.

The M3 and M3A1 without suppressor are indeed relatively accurate submachine guns. But they are no sniper weapons. Although the suppressor improved the accuracy, it was almost impossible to get a precise, targeted hit with a single shot—which of course would have been optimal for a silenced weapon. The demand for the development of a special suppressed submachine gun with high accuracy was often mentioned in the test reports but went unheard.

A special sheet-metal cap ("barrel thread protector") was used to protect the thread of the barrel nut while the weapon was not in use. *Courtesy of Corey A. Shaffer / Thompson Machine Company*

Comparison of the two types of the barrel thread protectors. The later one had a square, pressed-in recess for applying a tool for easier removal. *Courtesy of Corey A. Shaffer / Thompson Machine Company*

RIA-68-U268

UNCLASSIFIED

UNCLASSIFIED
Aberdeen Proving Ground
MARYLAND CONFIDENTIAL

A TEST OF BARREIS, SILENCED,
SUBMACHINE GUN M3, CALIBER .45 (U)

THIS DOCUMENT CONSISTS OF ..71..... PAGES
COPY ..10.... OF ..20.. COPIES, SERIES......
D.A. PROJECT NO. - 502-08-006
DEVELOPMENT AND PROOF SERVICES
FORTY-NINTH Report OCO Project No. TS2-2015
Regrading data cannot be predetermined

UNCLASSIFIED

In September 1956, some additional tests took place at Aberdeen, but no changes in design were taken.

Recon team leader Jim Bolen is firing his suppressed M3. Note the canvas heat protector covering the expansion chamber. *Courtesy of Jim Bolen*

Despite all efforts, the Bell suppressor was only about 80 percent as effective as the British counterpart of the Sten Mk. II S. This fact has been determined in a shooting competition in Aberdeen. Nevertheless, the suppressor remained unchanged and was successfully used in Korea, Vietnam, and other conflicts.

Highly decorated Vietnam veteran Jim Bolen used a suppressed M3 in his time with US Special Forces. He later wrote of his experiences in his book ***No Guts, No Glory.***

As a leader of a reconnaissance unit, he survived more than forty top-secret MACV-SOG (Military Assistance Command Vietnam, Studies & Observation Group) missions, which among others took him to Laos and Cambodia. He remembers the suppressed M3 very well:

Jim Bolen with reconnaissance team (RT) Auger, about to be choppered into Cambodia by the SOS for another secret mission. *Courtesy of Jim Bolen*

> Normally, one member of the recon team carried the M3 with an integral silencer. It was of .45 caliber and was used as a RON [remain overnight] gun in case at night an enemy, civilian, or even animal would accidentally happen on us; we could kill same without giving away our position or even the fact we were in the area, and for prisoner snatches to get in and out without being compromised.
>
> The weapon was reliable, but with the amount of ammo needed for a mission, it was very heavy. My gear weighed between 95 and 110 pounds depending on the mission. Weight was always a big problem on our missions, which was the reason we generally didn't carry large weapons such as the M-60, large amounts of claymores, or hand grenades on standard missions. When using the M3 on a prisoner snatch mission, the weapon is usually broken down and divided between all team members to distribute the weight, then assembled when needed.
>
> The M3 had one disadvantage in combat, yes. On more than one occasion when carrying the silenced M3, I would be confronted by NVA [the North Vietnamese army]. Normally I would have the advantage by opening up first on full auto, using my CAR-15, with every round being a tracer. This would scare the hell out of the enemy. But when I am carrying a silenced weapon, and I open up on them, they don't realize I am shooting at them and start returning fire ASAP. This is the truth and I will testify to it.

"This was late 1967 or early 1968, at Ban Me Thuot. I'd just got back from the range and was sitting on a bunker." Jim Bolen, author of *No Guts, No Glory*, has fond memories of the suppressed M3. *Courtesy of Jim Bolen*

> My team and I practiced a lot on the firing range with whatever weapon we were required to carry on a specific mission. To do this, we would take the silencer apart and clean the original screens with a wire brush and cleaning compound; once they become clogged, it reduces the silence capability. I never noticed any round being quieter than another. I know when we use the .223 round with a silencer, we have to use the first three rounds subsonic because of the crack sound of the high-velocity .223 round.

The High Standard Manufacturing Company also made a suppressed version of its small-caliber pistol "H-D" (.22) for the OSS at the same time. The design of the suppressor came from Bell Laboratories and was very similar to that of the M3, with its two chambers and the stainless-steel wire mesh parts. The gun was based on the suggestion of deputy director for research and development Stanley P. Lovell. From October 1943 until March 1944, 2,600 pieces with the designation "MS" (Military Silent) were delivered.

While fighting in Vietnam, a lot of foreign weapons fell into the hands of the Americans—including numerous Grease Guns that were made in China. Some of them were equipped with suppressors. However, since the OSS suppressors were not marked, it remains unclear whether the suppressors were originally made in the US or in Chinese workshops. An interesting finding is that the wire mesh cylinder in the expansion chamber was often replaced by oil-soaked felt in many of these suppressors.

USE BY THE IRISH REPUBLICAN ARMY

In the postwar years, the paramilitary Irish Republican Army (IRA) received most of its weapons from the United States and Libya. A secret network run by the Irish immigrant George Harrison organized and shipped larger quantities of arms and ammunition, especially for the "Border Campaign" (December 1956 to February 1962) to Ireland. These deliveries included M1 carbines as well as M3 Grease Guns, which were used by the IRA in attacks and robberies. Some of the M3s captured by the British had been equipped with suppressors. These devices were roughly manufactured home builds on which not even the small lateral holes in the barrel were deburred.

This M3 with homebuilt suppressor was confiscated by the Northern Irish police from a member of the IRA. *Courtesy of the chief constable of the Police Service of Northern Ireland*

The serial number has been ground off. *Courtesy of the chief constable of the Police Service of Northern Ireland*

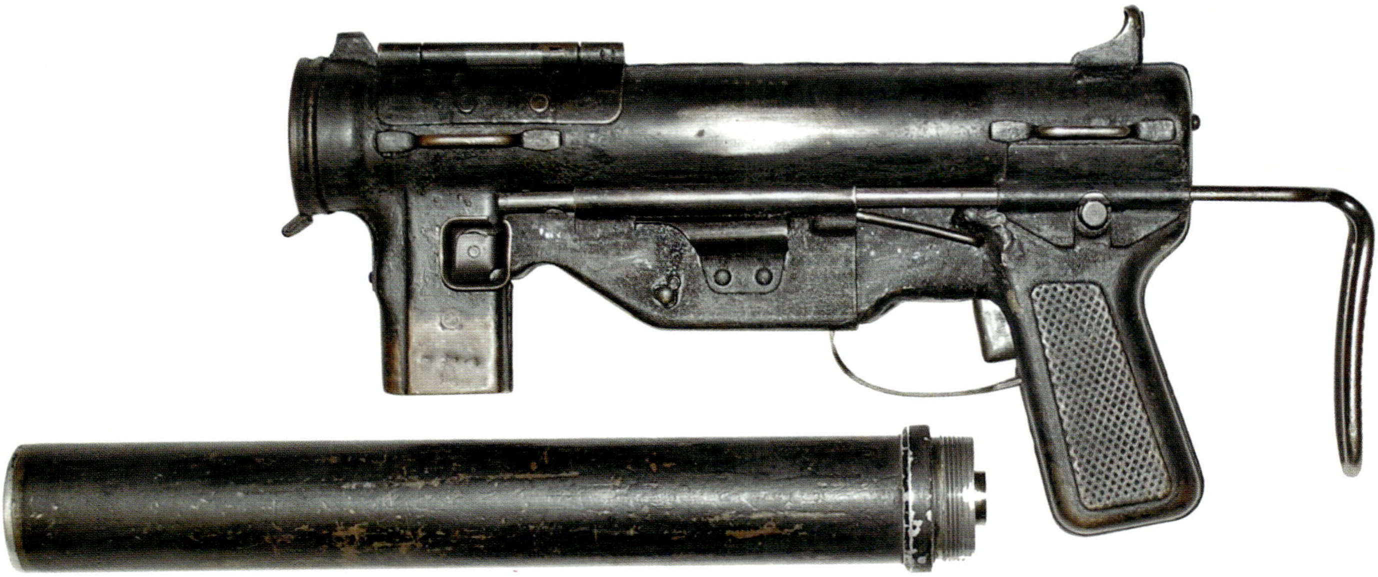

The barrel nut was modified similarly to the OSS version to hold the suppressor housing. *Courtesy of the chief constable of the Police Service of Northern Ireland*

Discovered IRA weapon cache with American weapons (M3A1 and Thompson M1921)

13 EXPERIMENTAL WEAPONS

DURING THE CURRENT SERIES production of the M3A1, no further changes in design were made. The weapon served its purpose, and the simplification potential seemed to be exhausted on this model. Nevertheless, one more elaborate experiment was carried out: the conversion to .30 carbine caliber (7.62 × 33 mm).

This model, with the designation "T-29," was built in the middle of 1944, with the aim of creating a fully automatic weapon using a cartridge that had more power and range than the current pistol caliber round used in submachine guns. At that time, the tests with the future model M3A1 were already in progress, and so this model was also used for this purpose instead of the somewhat dated M3.

The conversion of some preproduction M3A1 for the longer and more powerful cartridge in .30 carbine caliber was very costly: the magazine well, and magazine release button had to be adjusted to accommodate the carbine magazines of fifteen- and thirty-round capacity. The ejector was replaced by a spring-loaded one. The bolt had to be completely redesigned: around the bolt face, the material was machined off 3/8 inch, and the front guide rod retaining plate was increased in thickness the same amount (3/8 inch) as a counterpart. That way, the center part of the bolt face was completely enclosed by the reinforced plate when a round was fired. To make the bolt heavier, a steel block of ½-inch length was welded to its rear end, thus increasing its weight to 2 pounds, 4 ounces.

Problems were caused by the power of the carbine cartridge. The tension of the driving springs had to be increased—but not too strong; otherwise the operator would not be able to retract the bolt with his finger. A careful balance was needed. The problem was solved through the use of a short high-rate spring and a longer low-rate spring. The shorter spring was to keep the bolt from contacting and damaging the rear portion of the receiver. The longer spring was used to move the bolt forward, enabling the feeding and chambering of the cartridge.

A total of three M3A1 submachine guns were converted to T-29 configuration, and every weapon fired 1,500 rounds for test purposes. They functioned satisfactorily, but the muzzle flash and report of firing the .30 carbine round from the short submachine gun barrel got negative attention. Even the flash hider provided only little help. Nevertheless, the tests showed that more development work had to be done, such as adding a feed ramp for more-reliable operation. But it was not to be. At the same time, the development of the select-fire M2 carbine was successfully finalized, making any further work on the T-29 superfluous. On November 16, 1944, the project was officially terminated.

The Rock Island Arsenal keeps a heavily modified M3, serial number 0041925, in its collection. From the original weapon, only the receiver section with the magazine well, ejection port, and dustcover remained unaltered. The weapon can be cocked by a cocking handle on the right side. A plaque was probably once attached to the left side of the wooden stock. No further information about the weapon's purpose is known.

Experimental weapon T-29 (serial number X38) in .30 caliber. Note the longer barrel and the modified magazine well to take the smaller carbine magazines. *Courtesy of the Springfield Armory NHS / US NPS*

The bolt was also heavily modified. *Courtesy of the Springfield Armory NHS / US NPS*

The skeleton magazine was used only for testing purposes. *Courtesy of the Springfield Armory NHS / US NPS*

The second of three T-29s in .30 caliber (serial number X39). *Courtesy of the Springfield Armory NHS / US NPS*

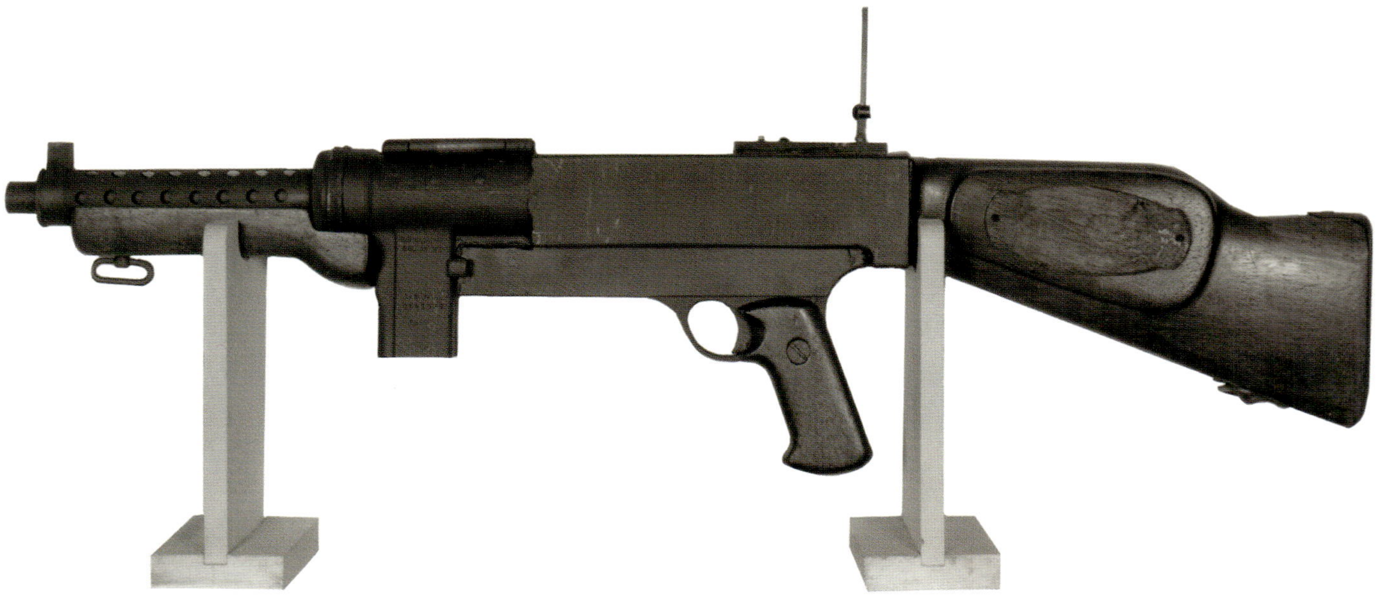

An extraordinary version of an M3A1 whose development history is unknown. *Courtesy of the Rock Island Arsenal Museum*

The purpose of this device attached to the barrel is not known. Maybe it was the test of a vehicle mount. *APG*

14 THE GREASE GUN IN COMBAT

THE M3 SAW ITS first combat action in the hands of American paratroopers who jumped off over France in the days before the D-day invasion in June 1944. Among the ground forces, the weapon was very scarce; it was handed out only to some Rangers and tank crews. In the further course of the war, however, increasing numbers of the M3 were shipped to Europe and distributed to the troops. But compared to other weapons, such as the Thompson, their numbers remained low, and the weapon is rarely seen in pictures.

It did not take long before some of the M3s fell into German hands. One of them was thoroughly examined by the Heereswaffenamt (German army ordnance) at the end of 1944. As can be clearly seen in the picture, the examined M3 has been converted for a cocking handle moving along a slot in the receiver. However, the report titled ***New American SMG "M3"*** does not mention this conversion. Probably the Germans at that time did not know that the M3 was usually cocked by a crank mechanism. The weapon got the German designation "MP 763 (a)" (a = American). Its design did not make much impression. The report concludes with the words "In its performance in combat the new American SMG is similar to the German MP40 and is therefore far inferior against the German MP44, which is an exceptional precision weapon firing a high-performance cartridge."

The number of M3 submachine guns that were delivered to resistance fighters and partisans in Europe is not known. It was all top secret, and most of the weapons remained (illegally) in private hands after the war. Even today, fully operational Stens, Grease Guns, and other arms are sometimes found in barns or during the demolition of old houses.

The M3 was used in much-larger numbers in the Pacific theater in the fight against the Japanese.

With the outbreak of the Korean War in 1950, the Grease Gun came back to the front line. Unlike the other infantry weapons, the Grease Gun could cope well with the extreme cold and the elements. The M1 Garand and the fully automatic M2 carbine became unpopular because of feeding problems. An official investigation had to deal with the complaint that even at short distances, the thick-clothed enemy soldiers did not fall despite several hits from the M2. With the Grease Gun, there were few climate-related problems—such as the somewhat painful cocking of the M3A1 with frozen fingers. All M3A1s used in Korea dated back to the year 1945. The start-up of an additional series production at the Ithaca Gun Company was delayed until the end of the war on July 27, 1953.

The last war where the Grease Gun saw action in large numbers was Vietnam. At that time, the main armament of the US troops had already changed to the modern M16 assault rifle (5.56 × 45 mm caliber) and its short version, CAR-15. Nevertheless, the Grease Gun remained popular because of its size and man-stopping power, especially in confined spaces such as in helicopters or armored vehicles. Large amounts were also delivered to the South Vietnamese troops and the US-friendly Montagnards (indigenous hill tribes).

And once again, yesterday's friends were tomorrow's enemies. China supported the North Vietnamese People's Army and the Viet Cong with weapons and ammunition, among them numerous Grease Guns and Thompsons that had been shipped as an auxiliary supply to China during World War II. Now they were in the hands of the enemy on the other side. In addition, the Viet Cong used captured Grease Guns that the US had once delivered to the French during the Indochina War.

After the Vietnam War, only a few special and training units kept the M3 and M3A1 in their inventory. In particular, the suppressed model proved still outstanding. And so, this model was also used by Delta Force for their prepared Operation Eagle Claw in Iran. This was an operation by the US armed forces to attempt to end the Iran hostage crisis by rescuing fifty-two embassy staff held captive at the embassy of the United States in Tehran on April 24, 1980. Iranian students had raided the embassy to force the extradition of the overthrown Shah Mohammad Reza Pahlavi from the US. The operation encountered many obstacles and was eventually aborted. As the US force prepared to leave, one of the Sea Stallion helicopters crashed into an EC-130 transport aircraft that contained both servicemen and jet fuel. The resulting fire destroyed both aircraft and killed eight servicemen. Operation Eagle Claw was one of Delta Force's first missions.

Some Grease Guns were still used during Operation Desert Storm, especially in armored vehicles.

Two German POWs of the 6th Armored Division are carried on a Jeep to the next camp near Brest, France, on August 28, 1944. *National Archives*

A group of American soldiers is marching through a town in Normandy for the press photographer. *National Archives*

For this young soldier, the war is over. One of the guards is equipped with an M3. *National Archives*

On the road in France with Jeep, hand grenades, and an M3. *National Archives*

Serving meals near Bouxwiller (Buchsweiler), in Alsace

Window shopping at a jewelry shop in Vielsalm (Bastogne district), Belgium, on October 15, 1944

"Stop for refreshments in St. Avold. Soldiers of Gen. Patton's 3rd US Army celebrate the capture of St. Avold, France, with wine." Today it is the largest American World War II cemetery in Europe, with 10,489 graves.

A soldier of the 325th Glider Infantry Regiment in Belgium takes cover behind a German *Sonderanhänger Ost 1,5t* (special trailer "East"), which is equipped with all-metal wheels because of a lack of rubber. *National Archives*

Soldiers of the 28th Infantry Division are marching to a new position during the siege of Bastogne on December 20, 1944. *Courtesy of the US Army Heritage & Education Center, Carlisle*

Soldiers of the 75th Infantry Division during the Battle of the Bulge, December 1944. *National Archives*

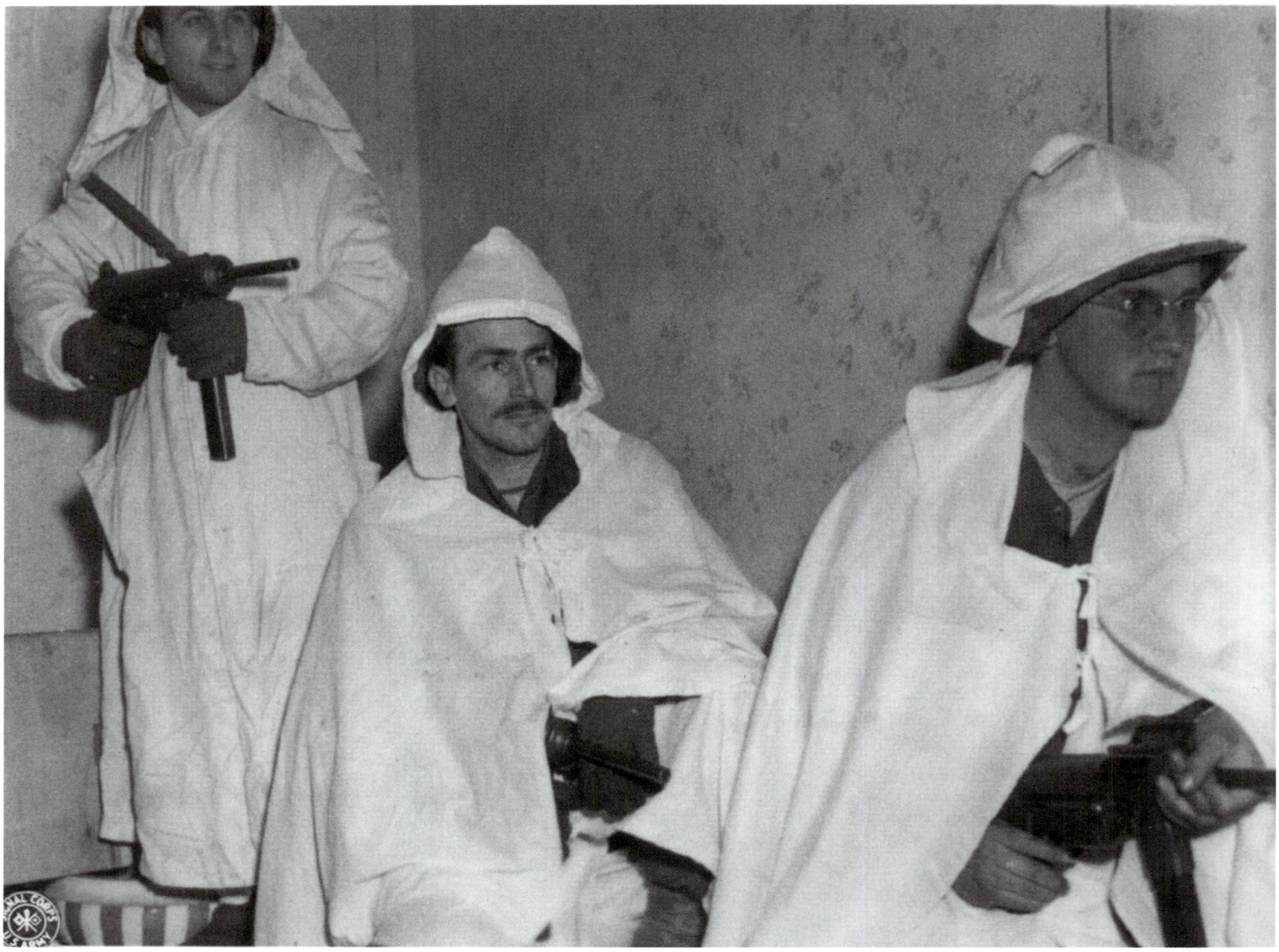

Three submachine gunners of the 104th Infantry Division, in winter camouflage, are ready for departure in the Ruhr area in March 1945. *Courtesy of the US Army Heritage & Education Center, Carlisle*

Soldiers of the 69th Infantry Division getting some rest, Germany, April 1945. *Courtesy of the US Army Heritage & Education Center, Carlisle*

The handy M3 was popular among tank crews. Here it is used by the 761st Tank Battalion "Black Panthers," a unit of African American soldiers. *National Archives*

Liberation of the POW camp (Stalag) on the German training ground of Hammelburg by the 14th Armoured Division. This "action photo" was arranged for press photographers. *National Archives*

Soldiers of the US Army and antiaircraft gunners of the Coast Guard during a passage. *National Archives*

"Come out Fighting" was the motto of Patton's 3rd Army's 761st Tank Battalion, which suffered a 50 percent casualty rate during World War II. Shown is Sgt. Harold Gary, commander of a 761st tank in Capt. Williams's A Company. He holds a tanker's "grease gun" and a .45-caliber automatic. The tank is a Sherman, used almost exclusively by the black armored unit. Miami News *photo*

On the advance through France. Note the M3 magazines taped together. The other soldier is equipped with a Browning Automatic Rifle (BAR). *National Archives*

Young Belgians are instructed in the handling of the M3 in September 1944. They should ensure public order in the liberated parts of the country. *Courtesy of the US Army Heritage & Education Center, Carlisle*

Wa A

Az 72 Wa Prüf 2 Stab
Bb.Nr. 180.11.44.

Berlin, den 16 November 1944
Anruf: J2 61 815

An

AHA/In 2

In 2
Eing. 19 NOV 1944
Nr. 11133/44

Betr.: Neue amk. MP "M 3".

Amerikanische MP "M 3". (Deutsche Benennung
MP 763(a)).

Kaliber:	11,43 mm	Visier:	Diopter-Stand-Vis.
Gewicht:	3,6 kg	Patr.Zuführung:	Magazin : 30 Patr.
Waffenlänge:	740/560 mm	Feuerfolge:	7–8 Schuß/sec.
Lauflänge:	200 mm	Kampfentfernung:	bis 200 m

Das Gehäuse der amk. MP "M3" ist in Blechprägetechnik hergestellt. Die Waffe hat einen Massenverschluß und ein doppelreihiges Stangenmagazin mit einreihigem Patronenaustritt und eine einschiebbare Schulterstütze. Waffentechnisch bringt die Waffe nichts Neues, denn sie besitzt den bekannten Massenverschluß, der in seiner Art schon bei der deutschen MP 18 und auch bei der

-2-

German report about a captured M3. Note the "straight-pull" cocking handle. The weapon got the German designation "MP 763(a)" (a = *amerikanisch*).

Kennblätter fremden Geräts D 50/1
Nordamerika

dtsch Kennummer

amk Ben:	Sub Maschin Gun Cal. .45" M 3	**763** 01 (a)
Abk:	(Brief Case' Gun)	
dtsch Ben: wo nötig erläutern mit	**Maschinenpistole 763 (a)** — amk M 3 —	
Abk:	MP 763 (a)	

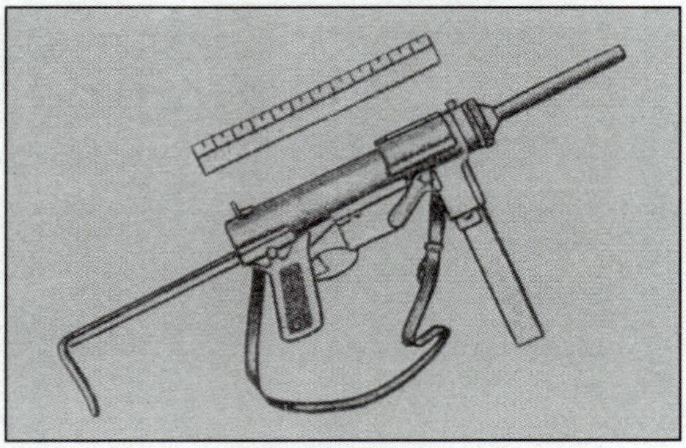

Kennzeichen

Bauart
Rückstoßlader mit feststehendem Lauf, Massenfederverschluß mit 2 Federn. Blechprägeausführung

Kaliber	11,43 mm
Gewicht	3,6 kg
Länge	740 mm / 560 mm
Lauflänge	200 mm
Zahl der Züge	8
Drallrichtung	rechts

Visier
Diopter Standvisier

Magazinart und Patronenzahl
Ansteckmagazin zu 30 Patronen

Patronen
Pist Patr „U. S. Army Kal. 45" (= 11,43 mm)

Feuergeschwindigkeit (theoretisch)
450 Schuß/Min

Gebrauchsentfernung
bis 200 m

The M3 was also added to the German summary D.50/1, containing description sheets of foreign equipment. At the time of issue, only a drawing of the M3 was available, and even in later reprints no photo was added.

Firing practice with the M3 in Burma in November 1944. *National Archives*

Marines searching a Japanese POW during the Battle of Okinawa, which lasted from April to June 1945

Souvenir photo with M3 and monkey in Asia

A group picture taken as a souvenir. The partisans are equipped with four M3s and one Beretta M1938A.

An American military policeman with his M3 guarding the border that divided Berlin until the reunification in 1990. The photo was taken in 1961, the year when the wall was built. *National Archives*

Maneuver of the US Army in Middle Franconia, Germany. George Anderson is posing with his M3A1 for a souvenir photo. *Courtesy of George Anderson*

Private First Class (PFC) Jose Ledoux-Garcia (Company C, 5th Battalion, 77th Armor) is sitting with his M3A1 on an M60A3 main battle tank during the NATO exercise "REFORGER" (Return of Forces to Germany) in 1985. *National Archives*

This crew member of an M24 also liked the handy M3. The size of the weapon was perfect for storing it in the limited interior space of tanks. The meaning of the number "5" written on the magazine well is not known.

Left: A member of the US Army is posing together with a German soldier carrying an M3A1 at the tank barracks in Sindelfingen on March 31, 1989. *Courtesy of R. D. I. Weigt*

Top: An American unit on the march through the Vietnamese bush and farmland. *National Archives*

The crew of an M24 "Chaffee" having a break during the Battle of Naktong Bulge, Korea, on August 17, 1950

Road construction work in Vietnam under American supervision. The soldier in the foreground carries an M3. *National Archives*

Left: The M3A1 in Vietnam, in a souvenir photo taken by Ken Hinks in 1967 as a reminder of his time there. *Courtesy of Ken Hinks*

Right: The "Montagnards" were native hill tribes in Vietnam, Cambodia, and Laos; they often supported the Americans in fighting the Vietcong. Note the unusual magazine pouch. *National Archives*

Montagnards during training with the M3A1. Note the attached flash hiders. *National Archives*

Weapons captured in the fight against the Vietcong: M3, PPSh41, MAT49, MG34, M60, BAR, RPD, and Mosin-Nagants. Some weapons changed sites several times.

A collection and recycling point for military scrap near Saigon. Between various weapons also some piles of M3s and M3A1s are waiting for the melting furnace. These photos were taken in 1990. *Courtesy of Mick Ranger & Philip Petersen / www.project-x.org.uk*

Still today the remnants of the war are present. This M3A1 was found on the Hồ Chí Minh trail near Ta Oy, Laos. Note the limiters that were attached to the stock. *Courtesy of Don Duvall / www.laosgpsmap.com*

Afghanistan 2010: After chasing away the Taliban from a settlement, the search for hidden booby traps begins. Here an M3A1 was dug in the sand but was found and recovered. Sgt. Ryan Robinson examines the weapon, which was slightly rusted but still in full working order. *Photos by Cpl. Matthew Narusberg, 144 Parachute Medical Squadron*

Afghan warriors handing over their weapons in the city of Herat in November 2010. Besides modern weapons an M3 can also be seen. *Courtesy of the US Army*

These weapons were confiscated by US forces in Iraq. Besides the Lee Enfield No. 4 Mk. I, several M3s and M3A1s are lying on the table. *Courtesy of the US Army*

An Afghan policeman with his M3A1. *Courtesy of the US Army*

15

USE IN FOREIGN COUNTRIES

AFTER THE END OF World War II, thousands of Grease Guns were scattered all over the world. They performed their service on both sides of the law. Thanks to its simple design, the weapon could be kept operational by any reasonably talented craftsman. Some countries even started to copy the Grease Gun, and wars such as in Korea or Vietnam once again brought large numbers of US stocks into the hands of allies and enemies. Many weapons disappeared into dark channels. The following is an overview of the most-important users of the Grease Gun after World War II.

CHINA

After China was increasingly weakened internally by the ongoing civil war since 1927 between Nationalists (Kuomintang) and Communists, Japan saw it as a welcome opportunity for expansion of its own supremacy in Asia and assurance of abundant natural resources in China. By the end of 1931, Japanese troops occupied Manchuria, and despite a ceasefire, there were repeated ground fighting and air raids. After a border incident on July 7, 1937, the second Sino-Japanese War began.

The Soviet Union supported the Chinese troops because they feared an expansion of Japanese influence. Even the anticommunist Nationalists received weapons and equipment. But China was also supported by the United States—they wanted not only to stop Japanese forces but also to win China as an ally for the period after the war and as a stable power in Asia. Under the Lend-Lease Agreement from February 1941, supply worth about 1.6 billion US dollars was delivered. In addition to the United Defense UD-42 submachine gun and various Thompson models, large numbers of Grease Guns were sent to China.

Many weapons had not been used by the warring factions in the fight but were stored away as a preparation for a final decision in the civil war. While the Kuomintang stubbornly fought against the Japanese and suffered heavy losses, the Communists avoided larger battles to save their strength for a foreseeable struggle against the Kuomintang. After the surrender of Japan, the civil war flared up again as expected. Given the escalating war, the United States conducted negotiations with the leaders of both sides, Mao Zedong (Communists), and Chiang Kai-shek (Nationalists), to find a peaceful solution. However, this plan failed. Thanks to further aid shipments from the United States, the Kuomintang could get Manchuria under their control before the fall of 1945, but eventually, they had to abandon the field in early 1949. They retreated with, their leader, Chiang Kai-shek, to the island of Formosa (today known as Taiwan), while on the mainland of China, the Communist leader, Mao Zedong, proclaimed the People's Republic of China on October 1, 1949.

In the years after the Japanese surrender, the Nationalists had also produced submachine guns and other weapons in their arms factories—primarily copies of the American Thompson and the M3A1. Neither weapon is easy to manufacture.

The Thompson requires extensive machinery for the milling procedures and well-trained workers, and the M3A1 requires various press molds, heavy presses, and a high degree of precision in the welding procedures to put the receiver halves together. Even Guide Lamp had had its difficulties. Where the machinery and molds originated and whether the United States had a hand in the setting up of the weapons manufacturing is not known. The assumption, however, is close, because the Chinese guns are exact copies of the American original. There are no deviations in design, and no simplifications or improvements. Even the checkered pattern of the pistol grip was copied exactly. Other weapons, such as the Russian submachine guns PPSh41 or PPS43, would have been much easier to manufacture.

A Chinese officer presents an M3 from an American supply shipment. *National Archives*

The Chinese markings betray the origin of this Thompson copy. *Courtesy of the National Museum of the Marine Corps, Quantico*

The Chinese even made Thompson copies in 7.62 × 15 mm Tokarev caliber. *Courtesy of the National Museum of the Marine Corps, Quantico*

Chinese soldiers with Thompson 1928 copies (recognizable by the typical front sling swivel)

一	二	三	四	五	六	七	八	九	十
1	2	3	4	5	6	7	8	9	10

Chinese numbers with translation

The first copy was the "model 36," so named for the year 1947 in the Chinese calendar. The weapons were manufactured at the 90th Arsenal in Shenyang, the capital of Manchuria in northeastern China. Until the Japanese occupation, the city was named Mukden. The markings on the magazine well bear the symbol of the 90th Arsenal: a "9" above a "0" in a circle, separated by a horizontal dash.

The production facility had an illustrious history. Once established by the Chinese general and later warlord Zhang Zuolin, the "Mukden Arsenal" produced ammunition for the Russian army in the occupied territories of Manchuria. After a great battle in the Russo-Japanese War, the Japanese were able to take the city on March 10, 1905. Although Manchuria was again returned to China, Japan secured its economic influence and built the South Manchuria Railway to Korea to transport raw materials more easily to its homeland. To protect the railway line, especially against Russian encroachments, Japan stationed the Kwantung army (also called Guandong army) in Manchuria. In September 1931, the "Mukden incident" (bombing attack) led to the complete occupation of Manchuria by Japanese troops and the establishment of the puppet state of Manchukuo under the ruler Pu Yi. After the outbreak of the second Sino-Japanese War in July 1937, the factory became part of the Japanese Nanman Arsenal and was renamed "Hoten Arsenal." From 1938 to 1945, the factory produced large quantities of ammunition for the Japanese forces. After the Japanese surrender, the Chinese Nationalists took over the Hoten Arsenal and named it the "90th Arsenal" in 1946.

For the civil war against the Communists, weapons were urgently needed. And so, in 1947, the production of the M3A1 copy began in its original .45 caliber as "model 36." Noticeable is the barrel nut, which was made in the style of the old M3, without the wrench flats, and the absence of the oil can inside the hollow pistol grip. The magazine well bears Chinese markings. Detailed information on the production period and the quantities produced could not be ascertained. According to Chinese sources, fewer than 10,000 weapons had been completed before the factory was closed. The production quality did not meet the level of the American original; a great deal of work was done by hand. During assembly, the individual parts were manually fitted to one another, making a subsequent replacement of defective parts not always possible without problems. When trying to exchange barrels of some weapons in museums, the author found out that the threads are not compatible, and the two holes in the receiver rear end for supporting the bolt's guide rods were drilled slightly differently on all weapons.

Bayonet with markings of the 90th Arsenal in Shenyang

Below the symbol of the 90th Arsenal is written:
model 36
11 mm submachine gun
00414

The first copy of the M3A1 was the model 36 in the original .45 caliber. This one is serial number 00414. *Courtesy of the National Firearms Centre / Royal Armories, Leeds UK*

Another model 36 (serial number 01788). The integrated oil can of the M3A1 in the pistol grip was not copied by the Chinese. *Courtesy of the National Museum of the Marine Corps, Quantico*

Below: The Nanjing Arsenal in 1865. Note the interesting rocket launcher.

For lack of ammunition in .45 caliber, the Nationalists were soon looking for an alternative—and produced the M3A1 from 1948 onward, in 9 × 19 mm caliber. The parts used for this, in particular the adapter insert into the magazine well, are very similar to the American-made 9 mm conversion kit. This kit had certainly served as a template. It is not known if the magazines were manufactured in China or if they were former British Sten magazines. In accordance with the Chinese calendar, the weapon was named "model 37." Its production was carried out in the "60th Jinling Arsenal" in Nanking (Nanjing), then the capital of the Republic of China. The magazine well is emblazoned by the marking of the Chinese military (bow and arrows in a gearwheel) above the symbol of the arsenal. Why the production had been moved to Nanking is not known. Early arms still bear the designation "model 36," but in a different writing style. The switchover to the "model 37" marking was done sometime during the series production.

Very early model 37 with serial number 000054. Note the characteristic stepped 9 mm barrel. The necessary adapter in the magazine well is missing.

Ammunition in .45 caliber was scarce, so China copied the 9 mm conversion kit. The resulting model was named "model 37." *Courtesy of the National Museum of the Marine Corps, Quantico*

Left: Ordnance marking above the symbol of the 205th Arsenal. The designation was still "model 36" (serial number 004127). *Courtesy of the National Museum of the Marine Corps, Quantico*

Right: This weapon with serial number 10033 was already marked with the new designation "model 37."

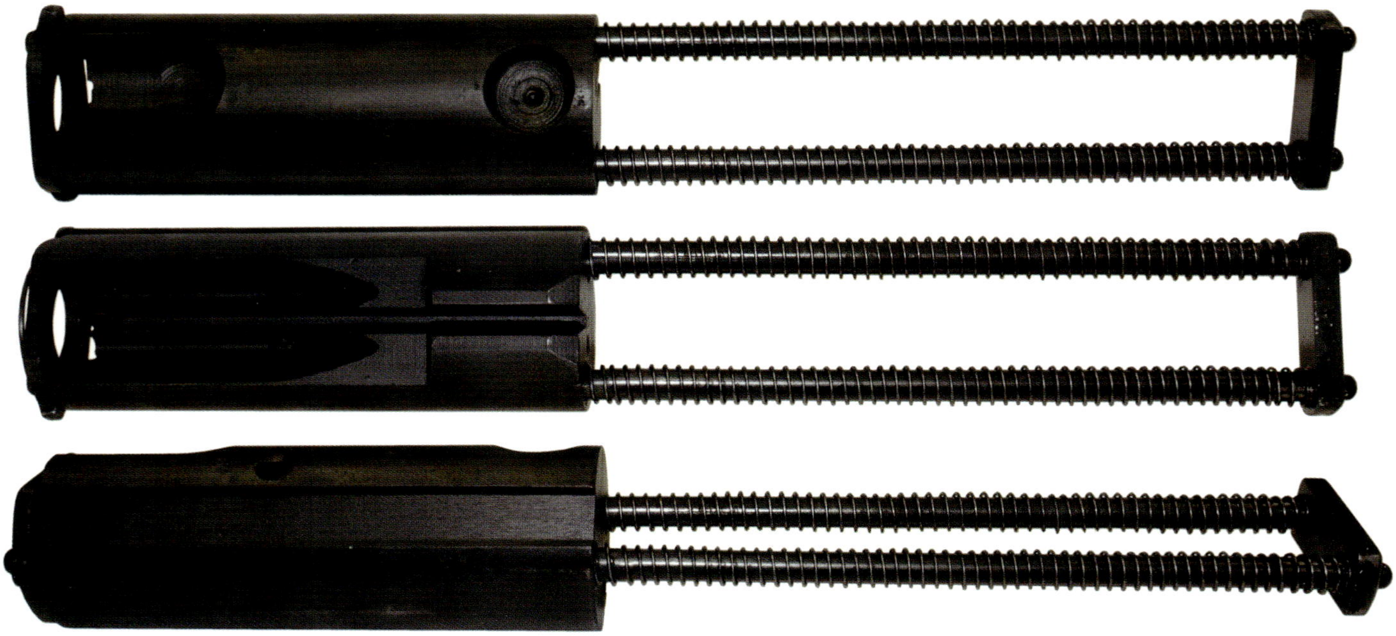

Bolt of the model 37 in 9 mm caliber. *Courtesy of the National Museum of the Marine Corps, Quantico*

In the further course of the civil war, the Nationalists could not stand against the Communists and at the beginning of 1949 had to retreat to the island of Formosa (today Taiwan / Republic of China). They also took the production documents and the necessary machinery of the 60th Arsenal from Nanking to set up a new production facility in Kaohsiung. Here the 205th Arsenal was founded while retaining its current arsenal symbol. On this occasion a rework of the weapon design was done: the magazine well was narrowed to accommodate the 9 mm magazines. A separate adapter was therefore no longer necessary, and the repeatedly recurring feeding failures belonged to the past. This change is clearly recognizable by a bulge in the side of the magazine well, since the shape of the Sten magazine is different than the M3/M3A1 magazines. In 1950, the first weapons, now called "model 39," could be completed on Formosa. As special accessories, a silencer and a curved barrel were available. The latter was rotatable and could be turned in any direction. A rubber sleeve served as a heat shield for the shooter's hand, so that he could grasp the barrel of the gun for better stabilization during shooting.

The 90th Arsenal in Shenyang in mainland China was taken over by the Communists and remained in operation. The former model 36 in .45 caliber was put back in production; the serial numbering started again. Nothing is known about the manufactured quantities. However, weapons from Communist production after 1949 are the rarest specimens of Chinese Grease Guns. They are recognizable by the different symbol of 90th Arsenal and the lack of a model designation on the magazine well.

Model 39 in 9 mm caliber (serial number 027057). The bulge in the side of the magazine well was necessary because of the different shape of the Sten magazine. *Courtesy of the ATF National Working Firearms Reference Collection*

Below the symbol of the 205th Arsenal the new designation "model 39" can be read. *Courtesy of the ATF National Working Firearms Reference Collection*

The rubber coating provides a safe grip and protects the operator's hand from the hot barrel. *Courtesy of the Republic of China Armed Forces Museum Taipei, Taiwan*

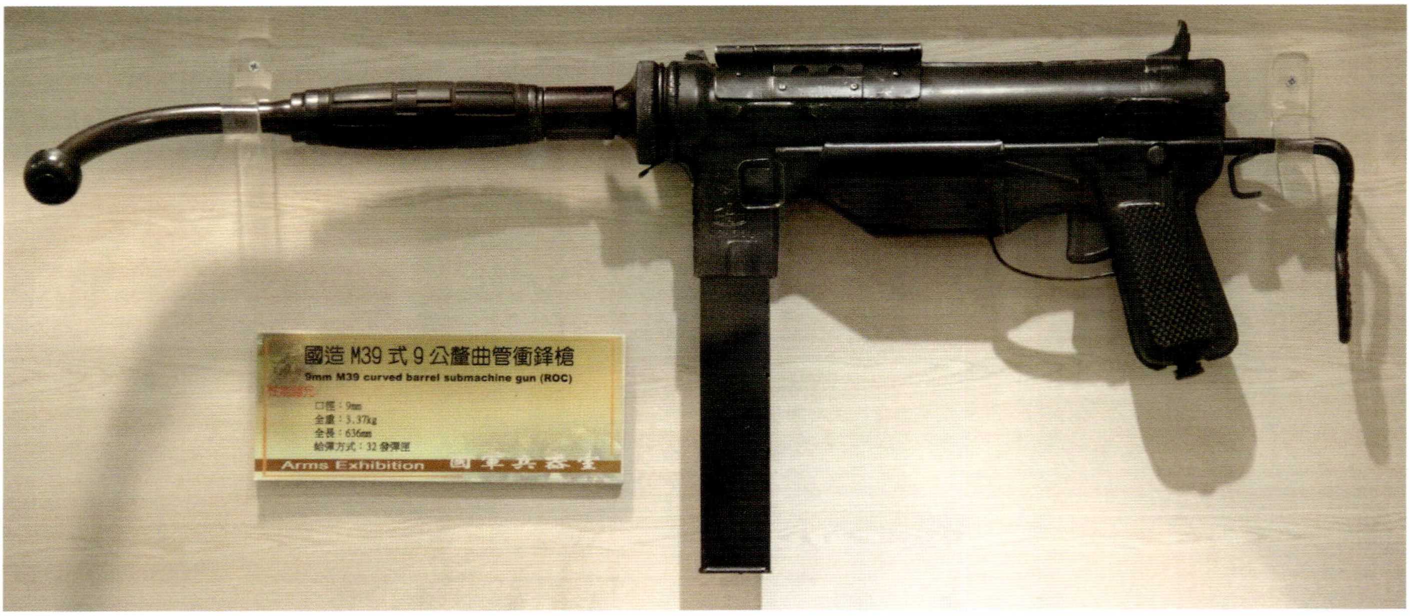

Model 39 made in Taiwan with a pivoted curved barrel. *Courtesy of the Republic of China Armed Forces Museum Taipei, Taiwan*

In 1949, the Communists resumed the series production of the model 36 (.45 caliber) in the 90th Arsenal. *Courtesy of Keregrant*

Marking of the weapon with serial number 028336. *Courtesy of the Republic of China Armed Forces Museum Taipei, Taiwan*

Changed symbol of the 90th Arsenal in Shenyang. No model designation was stamped on the magazine well. *Courtesy of Keregrant*

The new "Pistola Ametralladora P.A.M.1" in 9 mm caliber

In Vietnam, the US forces captured some Chinese "models 36," which were equipped with a suppressor. These might be copies or captured ones of US origin. Interestingly, in some cases the original stainless-steel wire mesh in the expansion chamber had been replaced by oil-soaked felt.

ARGENTINA

During World War II, Argentina was officially neutral. However, from the beginning of the war onward, the government was sympathetic with the German Reich and its allies. In 1943, the conservative government tried to change their course and get on the side of Germany's successful war opponents, but a military coup, with the participation of Juan Domingo Perón, prevented this change of direction. Only after massive pressure from the US government did Argentina declare war on the German Reich and Japan, on March 27, 1945. It was the last country in the world to do this.

The automatic pistol "Modelo Argentino 1927," made by FMAP DM. *Courtesy of the Rock Island Auction Company*

In June 1945, Argentina became a founding member of the United Nations (UN). A year later, Juan Perón won the presidential elections. Thanks to the forced industrialization, the country underwent an economic boom under his government, and a rising level of prosperity of the working class. At this time the Argentine military was already pursuing ambitious plans, especially those that provided the development of new weapons in their own country as well as the entry into nuclear technology. Like the victorious nations of World War II, Argentina also tried to attract German armament experts and scientists.

Even during the war, the Fábrica Militar de Armas Portátiles "Domingo Matheu" (FMAP DM), founded in 1936, had begun with the licensed production of the Colt M1911A1 automatic pistol in .45 caliber as the "Modelo Argentino 1927." In October 1942, the expanding factory was moved from Buenos Aires to Rosario, a town located about 186 miles (300 km) away in the province of Santa Fe. At the beginning of the 1950s, the company acquired additional rights for a licensed production of the American M3A1 submachine gun. No details are known. It was probably an initiative of the weapons factory to offer the Argentine military a domestically made submachine gun. A demonstration of the M3A1 for a commission of military personnel and civilians took place on October 21, 1950, at the premises of FMAP DM. The guest book of the company lists numerous participants, but without indicating their function and the reason for their invitation: Tte. Cnel. Pedro Alberto Merreins, Cap. Frag. José María Guzmán, Mayor Arnoldo Lutteral, Mayor Carlos Alberto Croce, Mayor Adolfo Ulises Senté, Capitán Jorge Granzinotti, Tte. de Navío Jorge F. D. Plater, Tte. 1° Humberto Juan Pizzi, Comandante Raúl A. Brollo, Sub.Of.May. Ricardo Gómez Costa, Of.Sub-Inspector Herminio M. Subleino, Sub.Comisario Nicolás Damis, and Gabino Castalli.

There was no support or information sharing for this project from the United States. Nevertheless, within a few weeks, the construction drawings could be completed under the direction of the Italian weapons designer Eduardo Sustercic from Cremona. The new weapon, in 9 × 19 mm caliber, was named "Pistola Ametralladora P.A.M.1." In contrast to China, the Argentine weapon has not been designed for using 9 mm Sten magazines. Instead, the magazine of the Grease Gun was copied and kept only slightly narrower. The weapon itself measures approximately seven-eighths of the original M3A1.

After the completion of the construction drawings, progress went on very slowly. Setting up the manufacturing facility proved to be troublesome. Shortages in qualified staff and space delayed the start of series production until 1954. This was caused mainly by the factory continuing the production of the Colt 1911 copy Modelo Argentino in 1927.

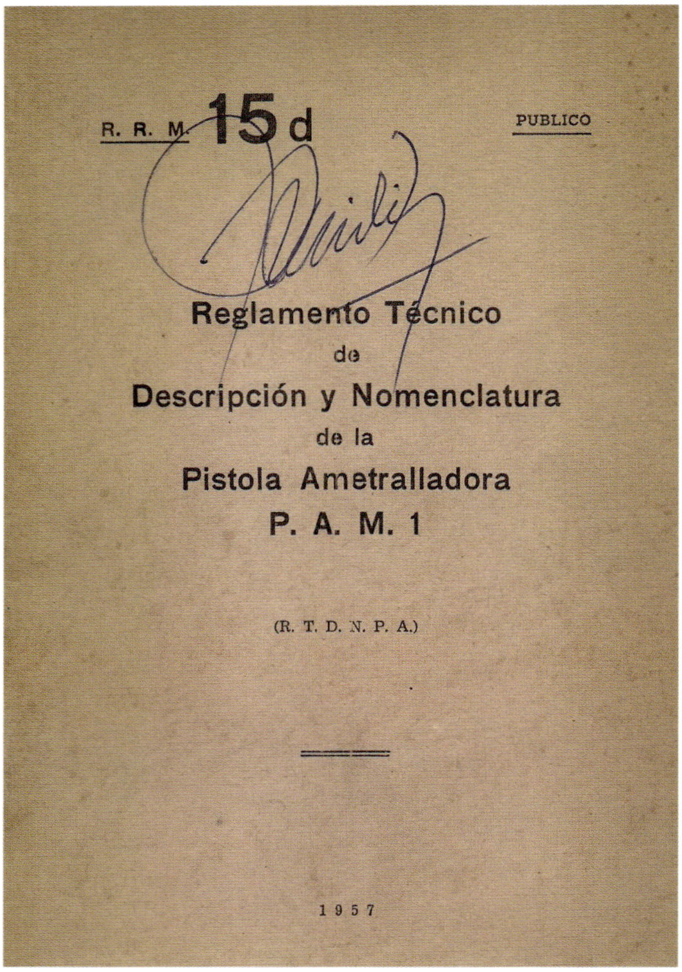
R. R. M. 15 d
PUBLICO
Reglamento Técnico
de
Descripción y Nomenclatura
de la
Pistola Ametralladora
P. A. M. 1
(R. T. D. N. P. A.)
1957

Cover of the manual *R.R.M. 15d* for the PAM 1, issued in 1957

Since the PAM 1 featured the same insufficient safety as its American example (blocking of the bolt by a hook-shaped nose attached to the dustcover), similar accidents were caused by dropping the weapon hard on the ground. The problem was solved by adding a grip safety behind the magazine well, similar to the Danish Madsen model MP 1950. The elongated locking lever reaches into the housing, where it is connected to a spring-loaded locking catch. Only after pressing the safety lever is the bolt released. This improved model was named "Pistola Ametralladora P.A.M.2."

So far, the following production figures and notes have been found:

1954: production of 5,000 PAM 1
1955: production of 3,500 PAM 1
1956: production of 6,500 PAM 1
1957: production of 5,621 PAM 1 (preparations for series production of the FAL assault rifle)
1958: production of 6,851 PAM 1
1959: production of 5,000 PAM 1
1960: production of 164 PAM 1
1961: production of 2,000 PAM 1
1962: no production of PAM 1 (begin of series production of the FAL assault rifle)
1963: conversion of 3,786 PAM 1 to PAM 2 (stored weapons from the arsenals)
1964: conversion of 5,292 PAM 1 to PAM 2 (stored weapons of the Navy)
1965: conversion of 5,754 PAM 1 to PAM 2 planned, but only 4,987 weapons were received and 4,504 converted
1966: conversion of 2,840 PAM 1 to PAM 2, plus additional production of conversion kits for 3,847 weapons
1967: no activity
1968: conversion of 66 PAM 1 to PAM 2
1969: production of 1,100 PAM 2
1970: end of production of PAM 2 and begin series production of the new "PA 3(DM)" submachine gun.
1971: no activity
1972: conversion of 56 PAM 1 to PAM 2

Like the M3A1, the stock of the PAM can be used as a tool to remove the barrel nut.

Markings on a PAM 1: D.G.F.M. means "Dirección General de Fabricaciones Militares," and F.M.A.P. means "Fábrica Militar de Armas Portátiles." *Courtesy of Eduardo J. Coppini*

The rear sight has two positions for 50 and 100 meters. *Courtesy of the National Firearms Centre / Royal Armories, Leeds UK*

The Argentine 9 mm barrel was not stepped like the Chinese one. *Courtesy of Eduardo J. Coppini*

Stock of the PAM 1 with integrated magazine-loading tool. *Courtesy of Eduardo J. Coppini*

Even the oil can in the pistol grip was copied from the M3A1. *Courtesy of Eduardo J. Coppini*

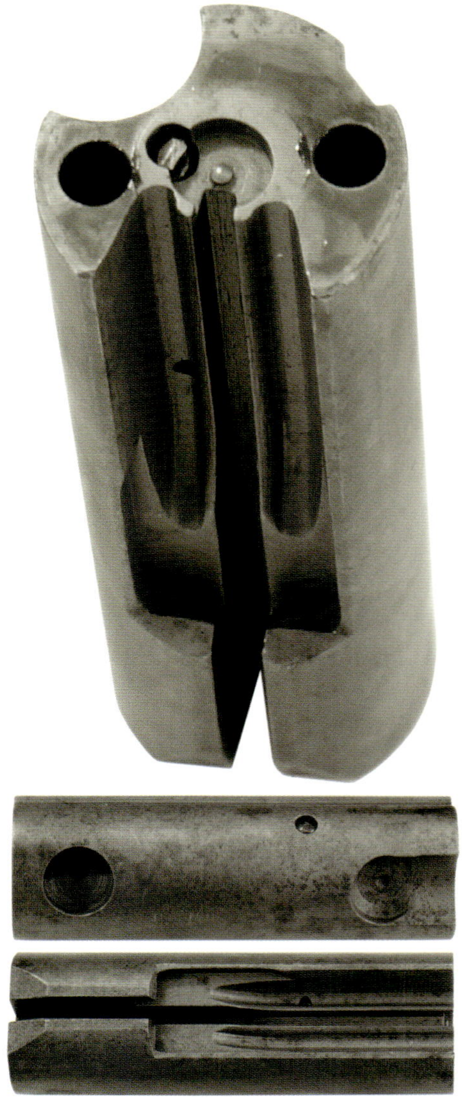

The bolts of the PAM 1 and 2 vary only slightly in the dimensions from the M3A1 bolt. *Courtesy of Eduardo J. Coppini*

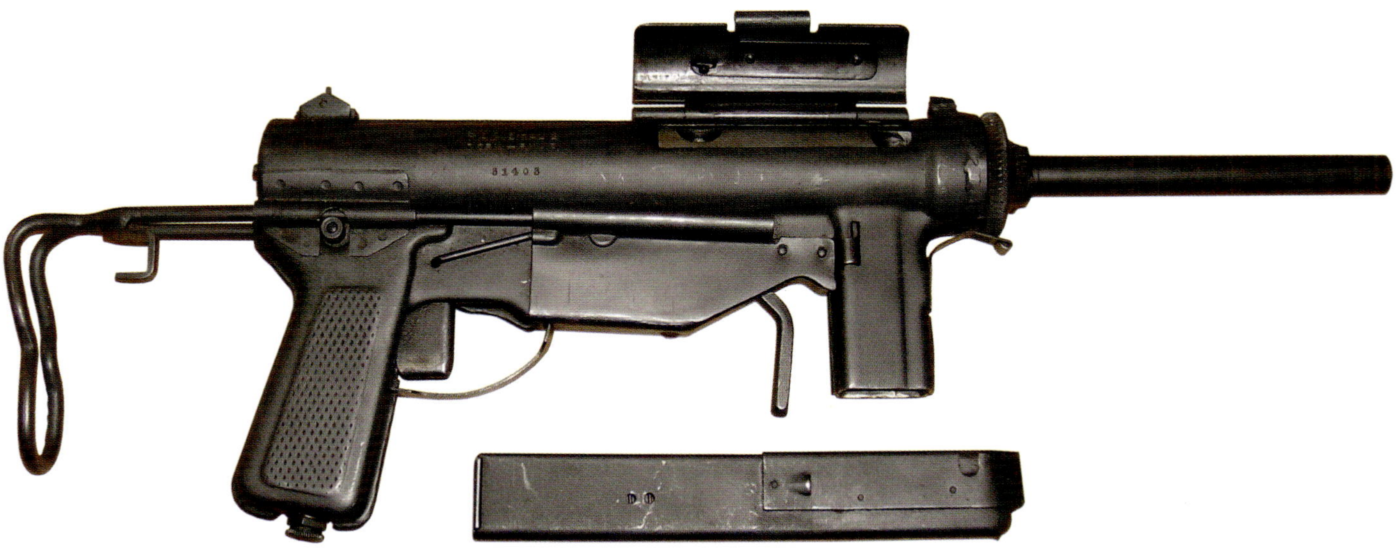

The simple safety mechanism with the hook blocking the way of the bolt was copied, but the Argentineans added an additional grip safety.

Like with the M3A1, after the barrel is unscrewed, the bolt of the PAM 1 and 2 can be easily removed. *Courtesy of the National Firearms Centre / Royal Armories, Leeds UK*

The additional grip safety was subsequently added to most of the PAM 1s. *Courtesy of Eduardo J. Coppini*

The grip safety is of simple design. An elongated locking lever reaches into the housing, where it is connected to a spring-loaded locking catch. Only after the safety lever is pressed is the bolt released. *Courtesy of Eduardo J. Coppini*

Traces of the modification work are clearly visible. *Courtesy of Eduardo J. Coppini*

Thus, the total production was at least 34,636 PAM 1s, of which 16,544 weapons were subsequently converted to the model PAM 2. Only 1,100 weapons left the production line originally configured as PAM 2 submachine guns.

The introduction of the submachine gun PA 3(DM) as the "FMK-3" ushered in the end of the PAM 2. This compact weapon was developed in the late 1960s by the Fábrica Militar de Armas Portátiles "Domingo Matheu" as a possible successor. With its magazine well integrated in the pistol grip and the tubular housing, it resembles the Czech submachine gun SA vz. 23. Approximately 30,000 weapons were produced until 1991. However, the PAM 2 remained in active service for many years and was also used in the battle of the Falkland Islands (Guerra de las Malvinas) from April to June 1982. In the 1990s, many PAMs were withdrawn from service and sold on the civilian market. Most of them were converted to semiautomatic because very few registered collectors were allowed to acquire a full-auto weapon.

Markings of a PAM 2

TECHNICAL DATA, PAM 1 AND PAM 2

Caliber	9 × 19 mm Parabellum
Weight (unloaded)	7.12 pounds
Length (folding stock, folded)	21.1 inches
Length (folding stock, extended)	28.5 inches
Barrel length	7.9 inches (6 grooves, right-hand twist)
Cyclic rate	450 rounds per minute
Sight	Folding leaf rear sight, 50 and 100 meters
Muzzle velocity	360 m/sec.

The accessories specifically for the PAM machine guns were limited to a strap, magazine pouches, and a carrying bag. The latter was used mainly by vehicle crews. A silencer was tested but not adopted.

Argentine soldiers line up with their PAM submachine guns.

Group picture taken on the Falkland Islands (Islas Malvinas). Most of the weapons are license-made FN FAL assault rifles. The PAM 2 is seldom found in photos.

Guard duty at a military camp. The weather of the Falkland Islands is very cool and windy throughout the year.

Souvenir photo with a PAM 2 after arrival on East Falkland (Isla Soledad)

Stick magazine of Argentine origin for thirty rounds. It is slimmer than the American magazine because of the smaller caliber (9 mm).

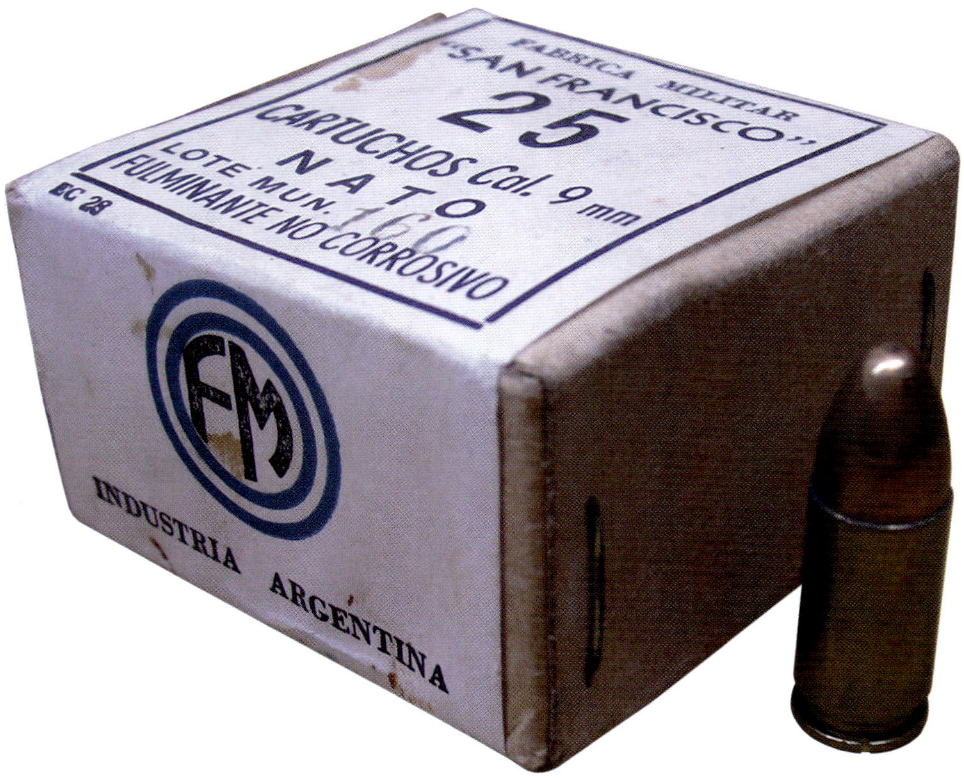

Headstamp with maker markings "FMSF" and 1982 date of manufacture. *Courtesy of Eduardo J. Coppini*

A package with twenty-five-round "Cartuchos Cal. 9 mm" made in Argentina by the FMSF (Fábrica Militar "San Francisco"). *Courtesy of Eduardo J. Coppini*

A dummy round (*right*) was made for training purposes. *Courtesy of Eduardo J. Coppini*

Most magazine pouches were green colored, but some were left untreated.

The early magazine pouch for the PAM was costly to manufacture.

The later, simplified model for two magazines. The lateral holes in the magazine were used to check the number of rounds left.

Very rare variant made of canvas with leather reinforcements. *Courtesy of Gaston Binnerbini*

Argentine soldiers with PAMs. Note the rare magazine pouches.

The protective cover was mainly used by vehicle crews. *Courtesy of Eduardo J. Coppini*

Some slightly different models of the covers were made. All of them were made of canvas, with leather reinforcements at critical wear points.

A suppressor was tested with the PAM but eventually never adopted. *Courtesy of the Museo de Armas de la Nación, Buenos Aires*

After their surrender, Argentine soldiers handing over their weapons to the British

Piles of Argentine weapons after the surrender. PAMs can be seen in the middle and at the right side. All others are mostly license-made FN FAL assault rifles.

A British soldier inspects a PAM 2 in front of a pile of former Argentine weapons.

Soldiers of the 3rd Greek Mountain Brigade "Rimini" marching in the streets of Athens in 1944. *Courtesy of the Athens War Museum*

Another picture of the 3rd Greek Mountain Brigade "Rimini" in 1944. *Courtesy of the Athens War Museum*

The police platoon of the 3rd Greek Mountain Brigade "Rimini" in 1944. *Courtesy of the Athens War Museum*

In the civil war, many World War II weapons such as the M3A1 served on both sides. *Courtesy of the Athens War Museum*

Exhibition about the Korea War at the Athens War Museum. *Courtesy of the Athens War Museum*

Patch of the Greek Expeditionary Force in Korea

Greek Expeditionary Force in 1951: morning line-up with M3s. *Courtesy of the Athens War Museum*

Greek Expeditionary Force in 1951: posing with the M3 for the photographer. *Courtesy of the Athens War Museum*

Greek infantry on exercise in 1966. This M3 is equipped with a flash hider. *Courtesy of the Athens War Museum*

GREECE

In Greece, the Grease Gun played an important role in the military. It remained in service for more than thirty years. The weapons found their way to Greece in the hands of the military units that were formed in 1944 by the Greek government in exile. These units got some of their equipment from the British and Americans before they could engage in the fighting in Europe. The 3rd Greek Mountain Brigade "Rimini" (Triti Elliniki Oriní Taxiarchía III Ελληνική Ορεινή Ταξιαρχία), for example, was formed on July 1, 1944, in Lebanon and arrived at Taranto in Italy by boat on August 11, 1944. There the unit was first under the command of the New Zealanders, and later in the Battle of Rimini under the command of the Canadians. After the city was taken, the unit received the title of honor "Rimini Brigade" (Taxiarchva Rvmini). By mid-October, the unit was still fighting in Italy, then the soldiers embarked and returned to Greece with full armament and equipment.

Although the German occupiers had left the country, the situation remained volatile. The establishment of a civilian government was difficult. The primary objective was to disarm the leftist guerrilla organizations National Liberation Front / EAM (Elliniko Apeleftherotiko Metopo), National Liberation Army / ELAS (Ethnikos Laikos Apeleftherotikos Stratos), and other armed organizations. But the controversy about the modalities of the disarming escalated and led to bloody conflicts in December 1944. In the heavy fighting of the so-called Battle of Athens, better known as ***Dekemvriana***, the Rimini Brigade was also involved. But only after the intervention of British troops in January 1945 could the situation be brought under control again. At that time, 75,000 British soldiers were positioned in and around Athens. Around 7,000 people on both sides had lost their lives in these thirty-three eventful days of fighting.

Sadly, the peace did not last long. The still-smoldering conflict between left- and right-wing groups flared up again in March 1946 and led to three and a half years of civil war. While Albania, Yugoslavia, and the Soviet Union supported the People's Front and their "Democratic Army of Greece," the conservative government got aid from Britain and the United States. Besides the Grease Gun, many other weapons saw action on the side of government forces, especially the submachine gun models Thompson M1928 and M1A1 and Sten Mk. II and Mk. III. The Greek navy received a supply of British Lanchesters. Political changes, in particular the break of the Yugoslav leader Tito with Stalin, ensured a strong decline in support for the leftist forces. Thanks to the support from the West, the Greek government forces could eventually gain the upper hand, and on October 9, 1949, the official cessation of hostilities was announced. The civil war was fought with the utmost severity and devastated the country, followed by emigration, expulsion, and oppression of ethnic minorities.

The M3 and M3A1 remained in the arsenals of the Greek army and were issued to the noncommissioned officers of the Greek Expeditionary Force (GEF); they came to Korea in November 1950. From January 1951 onward, up to 2,000 Greek soldiers fought against the Chinese army before they were ordered back after a conflict with the NATO partner Turkey in mid-1955. Until the 1970s, the Grease Gun remained in service with the paratroopers and special forces. In 1973, their withdrawal from service was decided. The good old Grease Gun was then gradually replaced by various versions of the FAL assault rifle.

NORTH KOREA

By 1948, as a product of the Cold War between the Soviet Union and the United States, Korea was split into two regions, with separate governments. Both governments claimed to be the legitimate government of Korea, and neither side accepted the border as permanent. The civil war escalated into open warfare when North Korean forces (supported by the Soviet Union and China) moved to the south to unite the country on June 25, 1950. South Korea requested support from the UN. On June 30, 1950, President Truman deployed US ground forces to Korea and a naval blockade of North Korea. At that time, the US infantry was equipped with the M1 Garand, M1 and M2 carbines, and the Grease Gun (mostly model M3). During their advance to the south, the North Koreans captured a large number of American-made weapons.

Even today, an array of weapons is used by the North Korean troops. In parades, one can see soldiers that are still equipped with Russian PPSh41 and PPS43 or, indeed, Grease Guns. However, probably not all of them are in functional condition. Referring to the M3, the crank-style cocking handles are in all sorts of positions or even lacking. Sometimes even the entire cocking handle assembly and the trigger are missing.

Some of the unserviceable weapons ended up in the Victorious Fatherland Liberation War Museum in Pyongyang, where colorful paintings show the great victory of the North Korean troops over the imperialist invaders. Large equipment dominates the exhibition of captured weapons, but alongside the stacks of carbines and Thompsons are M3 and M3A1 submachine guns.

Parade with Grease Guns in Pyongyang. Note the varying positions of the crank-style cocking handle. Presumably most of the weapons are not in working condition anymore.

REPUBLIC OF THE PHILIPPINES

The expansion of the Philippine marine corps (Hukbóng Cawal Pandagat ng Pilipinas) and navy (Hukbóng Dagat ng Pilipinas) since the year 2000, due to the increased global terrorist threat, resulted in an increased need for compact small arms. However, only a low budget was available for the implementation of these plans, and so the acquisition of modern weapons was very limited. As a temporary remedy, the Navy Office of Armament & Communications (N-11) decided to release several hundred M3s that were stored as a reserve. The weapons were thoroughly cleaned and repaired if necessary. Some of them were modified by adding Picatinny rails and Simmons red-dot sights. For the price of one original Heckler & Koch UMP in .45 caliber, about forty Grease Guns could be reworked.

A suppressed version of the M3 was developed in a local workshop for the Special Operations Platoon of the Marine Corps. Information about the inner design of the suppressor is not available. The first prototype was presented in May 2004 at the Marine base of Fort Bonifacio and performed well on the shooting range. A small series production in an unknown quantity followed.

A few M3s were carefully nickel-plated and issued to the guards of the Navy headquarters on Roxas Boulevard in Manila.

AUSTRIA

Five years after World War II, the "Bereitschaftsgendarmerie (B-Gendarmerie)" (riot-control police) was founded in the American sector of Austria. Unlike the regular police, which was armed with the old M95 carbine, the B-Gendarmerie received light infantry weapons from US stock; mostly M1 carbines and M1 Garands with bayonets and rifle grenade launchers. At the end of 1953, the B-Gendarmerie consisted of about 4,000 men. Thanks to the existence of this organization, it was possible to establish the new Bundesheer (army) on May 15, 1955. Its armament came from Allied depots.

Besides Russia (Mosin-Nagant M44, PPSh41) and France (MG34, MG42, K.98k), the United States gave 24,500 M1 carbines, 500 M2 carbines, 25,000 M1 Garands, 1,837 self-loading pistols 1911A1, 623 M3s, and 1,885 M1A1 Thompson submachine guns.

For consistent training and handling, instructions (ZI.96.363-3/57) were issued by the BMI on July 25, 1957.

The American submachine guns remained in service for more than ten years before they were gradually replaced from 1967 onward by the UZI.

Painting in the North Korean Victorious Fatherland Liberation War Museum: "The United States will have to pay the price for what they have done." *Courtesy of Lindsay T. Fincher*

The Grease Gun had no chance against the North Korean superiority. Oversized painting in the Victorious Fatherland Liberation War Museum. *Courtesy of Lindsay T. Fincher*

Piles of captured weapons are proudly presented. Here, Grease Guns and Thompsons are surrounded by Bren machine guns, M1917 rifles, Colts 1911, and various other equipment. *Courtesy of Lindsay T. Fincher*

The M3 in service with the Naval Special Operations Group (NAVSOG), also known as Philippine Navy SEALs. *Courtesy of MBLT6 / N-11 PN(MC)*

Above: Markings on a reworked M3: "PHIL. NAVY SUPP M3" and serial number "016." *Courtesy of N-11 / Philippine Government Arsenal*

Right: A row of freshly reworked M3s with suppressors waiting for the test firing. *Courtesy of N-11 / Philippine Government Arsenal*

A reworked M3 with suppressor ready for delivery. Note the camouflage pattern color and the modern aiming device. *Courtesy of N-11 / Philippine Government Arsenal*

Weapons exhibition of the Austrian Bundesheere in 1957, at the military training area Allentsteig: MP M1 (Thompson), MP M3 (Grease Gun), MP40, MP41 (PPSch41), M1 carbine, Karabiner 98k. *Courtesy of the Federal Ministry of the Interior, Vienna, Austria*

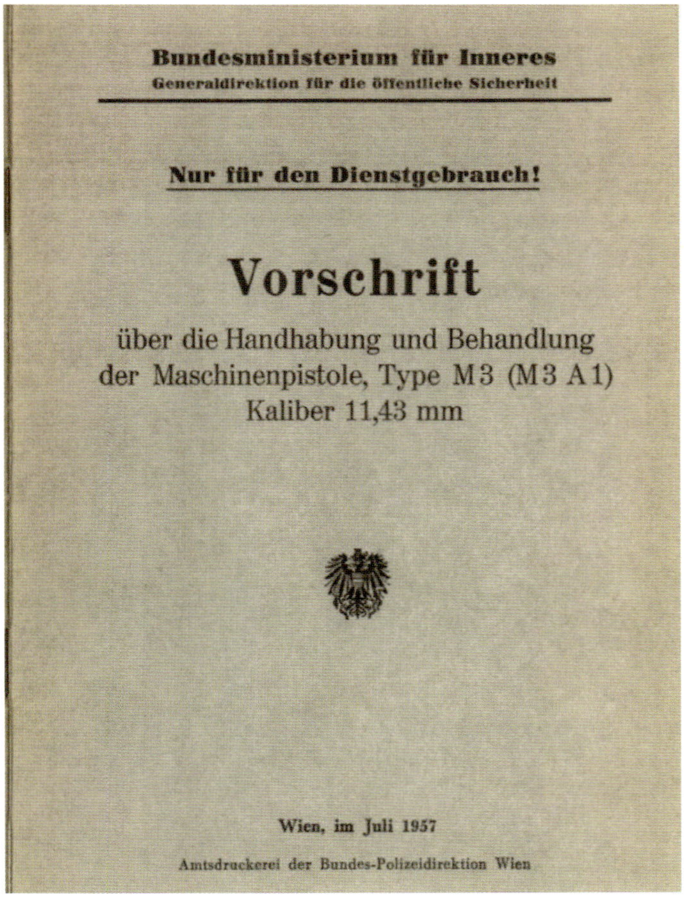

Bundesministerium für Inneres
Generaldirektion für die öffentliche Sicherheit

Nur für den Dienstgebrauch!

Vorschrift

über die Handhabung und Behandlung der Maschinenpistole, Type M3 (M3 A1) Kaliber 11,43 mm

Wien, im Juli 1957

Amtsdruckerei der Bundes-Polizeidirektion Wien

Service regulation Zl.96.363-3/57 for the police, issued by the Austrian Federal Ministry of the Interior in July 1957. *Courtesy of the Federal Ministry of the Interior, Vienna, Austria*

I) Technische Daten der MP 3 und der Normalpatrone

Kaliber	11,43 mm
Länge des Laufes	200 mm
Anzahl der Züge	4
Drallrichtung	konstanter Rechtsdrall
Länge der MP 3	57,9 cm
Länge der MP 3 mit ausgezogener Schulterstütze	75,6 cm
Gewicht mit leerem Magazin, Öler und Riemen	4,04 kg
Gewicht ohne Magazin	3,65 kg
Gewicht des leeren Magazins	0,35 kg
Gewicht mit vollem Magazin	zirka 5,00 kg
Gewicht der Normalpatrone	20,1 g
Gewicht des Geschosses	15,2 g
Gewicht der Treibladung	0,32 g
Anfangsgeschwindigkeit des Geschosses	zirka 270 m/sek.
Mündungsenergie	49,5 m/kg
Höchstschußweite	zirka 1.500 m
Das Maximum der Geschoßwirkung liegt zwischen	180—200 m
Feuergeschwindigkeit	350—450 Schuß/min.
Streuung bei aufgelegter MP 3 auf	
25 m = zirka	22 cm mittlerer Radius
50 m = zirka	36 cm „ „
70 m = zirka	58 cm „ „
Aufsatzstellung (unveränderlich)	100 Yard = 92 m.

Technical data of the M3 according to the service regulation. *Courtesy of the Federal Ministry of the Interior, Vienna, Austria*

Die Maschinenpistole M 3 (M 3 A 1), Kaliber 11,43 mm
(Einzelteile)

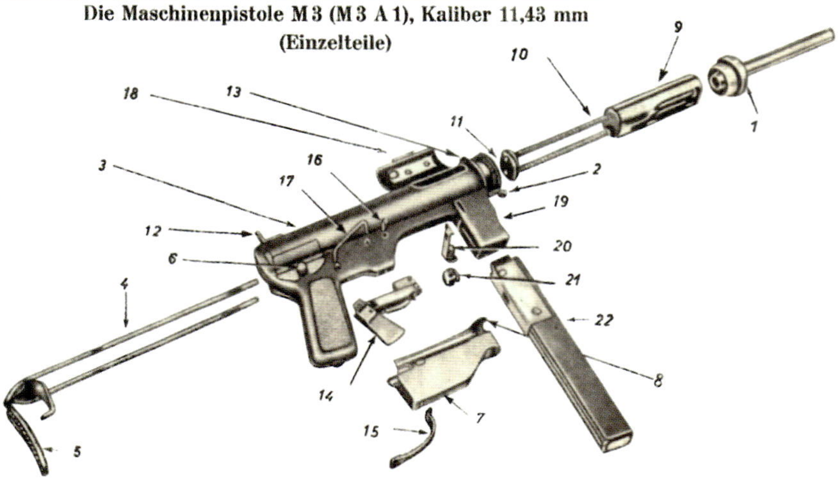

Verzeichnis der Einzelteile:

1 Lauf
2 Laufsperrhebel
3 Gehäuse mit Griffstück
4 Schulterstütze
5 Auflage zur Schulterstütze
6 Fixierbolzen zur Schulterstütze
7 Spannvorrichtung mit Gehäuse
8 Auswerfer
9 Verschlußstück
10 Führungsstangen mit Vorholfedern
11 Führungsstangenstützplatte
12 Ring- oder Dioptervisier
13 Korn
14 Abzug, Abzugshebel, Abzugsstollen und -feder
15 Abzugsbügel
16 Abzugsstollenstift
17 Abzugshalter
18 Deckel mit Sicherungsnase
19 Kasten zur Magazinsaufnahme
20 Magazinshalter mit Feder
21 Magazinshalterschild (Schutzblech)
22 Magazin

Illustration of the disassembled M3 from Zl.96.363-3/57. *Courtesy of the Federal Ministry of the Interior, Vienna, Austria*

Line up of a Fahrenheit Kärnten platoon of the B-Gendarmerie. The unit is equipped with American M3 submachine guns, Jeeps, and M8 "Greyhound" light armored cars. *Courtesy of the Federal Ministry of the Interior, Vienna, Austria*

Minister of the interior Oskar Helmer visits a mechanized platoon of the new Bundesheer. The soldier on the left carries an M3. *Courtesy of the Federal Ministry of the Interior, Vienna, Austria*

16 FINAL WORDS

THE "GREASE GUN," WITH all its different variants and all the accessories, is an interesting field for collectors. Quite wrongly the weapon is overshadowed by the famous Thompsons. Both weapons have their own appeal: on the one side, the elaborate design and the high-quality workmanship of the Thompson, and on the other side, the uncompromising simplified sheet-metal weapon from Guide Lamp.

Shooting the "Grease Gun" is a lot of fun. Thanks to the low rate of fire it is easy to hit the target, and the muzzle does not rise up even when firing longer bursts. And thanks to the widespread availability of .45 ACP caliber, ammunition is not a problem.

So why not visit one of the well-known machine gun shoots in the United States or other countries? Some fun firing the good old "Grease Gun" is highly recommended.

For sale at Knob Creek: an M3 with flash hider, lying on the table between a G3 and a Steyr AUG

Fun with the M3A1 at the Knob Creek Gun Range in Louisville, Kentucky. *www.knobcreekrange.com*

And for the young enthusiasts, the Grease Gun is already available as deceptively real-looking airsoft models.